Agriculture, Diversification, and Gender in Rural Africa

Agriculture, Diversification, and Gender in Rural Africa

Longitudinal Perspectives from Six Countries

Edited by
Agnes Andersson Djurfeldt, Fred Mawunyo Dzanku, and Aida Cuthbert Isinika

OXFORD
UNIVERSITY PRESS

OXFORD
UNIVERSITY PRESS

Great Clarendon Street, Oxford, OX2 6DP,
United Kingdom

Oxford University Press is a department of the University of Oxford.
It furthers the University's objective of excellence in research, scholarship,
and education by publishing worldwide. Oxford is a registered trade mark of
Oxford University Press in the UK and in certain other countries

First Edition published in 2018

Published in the United States of America by Oxford University Press
198 Madison Avenue, New York, NY 10016, United States of America

British Library Cataloguing in Publication Data
Data available

Library of Congress Control Number: 2017945146

ISBN 978–0–19–879928–3 (hbk.)
978–0–19–879929–0 (pbk.)

Acknowledgements

This book is the result of a collective African-Swedish research effort involving many people over close to two decades.

The data presented have been collected by country teams in the six countries covered by the book, most of which appear in the following chapters. In addition to these, the Mozambique team was headed by Peter Coughlin and the Zambian team consisted of Mukata Wamulume and Audrey Kalindi. During the course of the project, data collection in three of the countries that were part of the earlier phases was not possible to fund during the final phase. The importance of these teams in the original study design should be acknowledged. Here the Uganda team was headed by Bernard Bashasha, the Nigerian team by Tunji Akande, and the Ethiopian team by Wolday Amha.

In Lund, apart from the team members who contributed directly to this book, we should acknowledge the input of Jakub Kocanda and Will English in transforming a messy Word document into a survey and a database that was instrumental in minimizing data entry entries. Maria Francisca Archila Bustos (although co-author of two chapters) deserves a special mention as the guardian of this database, dealing with numerous time-consuming tasks to get it into shape. Several of the chapters were presented at the Rural Development Seminar co-hosted by the Departments of Economic History and Human Geography. The input from Björn Holmquist and Sultana Nasrin is gratefully acknowledged in this context. Henrik Steen at the Faculty of Social Sciences has patiently and efficiently dealt with the many financial transactions from Sweden to the six country teams over several years. The inputs of Sophie Hydén-Picasso and Kjell Josefsson at Research Services, Lund University is also much appreciated.

The funding streams of Afrint III have lead to unexpected and highly rewarding new partnerships, which have contributed in different ways to the final product. Aside from the original funding from Sida—who generously paid for the data collection for the first and second rounds—funding for the third phase has been awarded from Sida's research council (U-Forsk), Vetenskapsrådet and the UK's Economic and Social Research Council (ESRC), and UK aid from the UK government. Afrint III is part of two grant schemes, DEGRP (DFID-ESRC Growth Research Programme) jointly funded by the UK's Department for International Development (DFID) and ESRC, and APRA (Agricultural Policy Research in Africa) conducted with funding from UK aid from the UK government. The findings and conclusions contained are those of the authors and do not necessarily reflect positions or policies of the UK government, DFID, or ESRC, or of the Swedish government, Vetenskapsrådet, or Sida.

While the funding from all of these sources is gratefully acknowledged, the DEGRP II and APRA have also provided exciting opportunities for intellectual exchange. In this context John Tompson, Ephraim Chirwa, Christine Okali, Rachel Sabates-Wheeler, Jim Sumberg, Ruth Meinzen-Dick, Dan Brockington, and Phil Woodhouse should be acknowledged. The opportunity to present work from Afrint in these contexts has shaped the book in concrete ways. In particular, Elizabeth Harrison's suggestion to include a chapter comparing male- and female-headed households led to a chapter on assets. Likewise, Steve Wiggins' review of Chapter 4 improved the cross-sectional analysis considerably. The support of Louise Shaxton and Natalie Brighty at the Overseas Development Institute and Oliver Burch, Beatrice Ouma and Lesley White at APRA is also appreciated.

Afrint III has benefited greatly from its three advisors, Sylvia Chant at the Department of Geography and Environment, London School of Economics, who has generously taken her time to discuss the project and provide advice. Steven Haggblade at the Department of Agricultural, Food and Resource Economics, Michigan State University, a long-time friend of Afrint, has contributed in numerous ways. Finally, Margaret Kroma at the World Agroforestry Centre has tapped into her rich stakeholder experience to provide this perspective on the project.

At Oxford University Press, Adam Swallow and Katie Bishop have guided us through the publication process with friendliness and professionalism.

Contents

List of Figures

List of Tables

List of Abbreviations

ADMARC	Agricultural Development and Marketing Corporation
APIP	Agricultural Productivity Investment Programme
ASWAp	Agricultural Sector-Wide Approach
CFE	control function estimates
CFU	Conservation Farming Unit
DFZ	Deciduous Forest zone
ER	Eastern Region
FAO	Food and Agricultural Organization
FASDEP	Food and Agriculture Sector Development Policy
FE	fixed effect
FIDP	Farm Income Diversification Programme
FISP	Farmers Input Support Programme
FFM	female farm manager
FMF	female-managed farm
FRA	Food Reserve Agency
GDP	gross domestic product
GSZ	Guinea Savannah zone
IGLS	Iterative Generalized Least Squares
IRLAD	Irrigation, Rural Livelihoods and Agricultural Development
LMI	lower middle income
MGDS	Malawi Growth and Development Strategy
MFM	male farm manager
MMF	male-managed farm
NAIP	National Agricultural Investment Plan
NAIVS	National Agricultural Input Voucher Scheme
NGO	non-governmental organization
NFI	non-farm income
PPP	purchasing power parity

RE	random effect
RGDP	real gross domestic product
RNFE	rural non-farm employment
SSA	sub-Saharan Africa
ST	structural transformation
TIP	Targeted Input Programme
UER	Upper East Region
USD	US dollar

List of Contributors

Sarah Alobo Loison is a PhD candidate (co-tutelle) in Social and Economic Geography at the Department of Human Geography, Lund University, Sweden, and in Economics at Montpellier SupAgro/Ecole Doctorale Economie et Gestion de Montpellier (EDEG) in conjunction with UMR MOISA, CIRAD, Montpellier Cedex 5, France. Her research focus is on household economics, structural transformation, and rural livelihoods and diversification.

Agnes Andersson Djurfeldt is the team leader of the Afrint group. She is a Professor of Social and Economic Geography at the Department of Human Geography, Lund University. Agnes' research interests focus on rural-based processes of transformation within and outside agriculture in sub-Saharan Africa involving changing rural and multi-local livelihoods.

Maria Francisca Archila Bustos is a Research Assistant at the Department of Human Geography at Lund University. She specializes in geographic information systems (GIS), spatial analysis, and remote sensing.

Göran Djurfeldt is Professor Emeritus in Sociology from Lund University. He is a veteran development researcher who has worked and written extensively on agricultural change and rural development in India and in sub-Saharan Africa. Göran was the team leader of the first and second phases of Afrint, and is currently working at the Department of Human Geography at Lund University.

Fred Mawunyo Dzanku has thirteen years of research experience and interests in the areas of the economics of rural households, applied econometric modelling, and agricultural production economics. He has extensive experience in the implementation of household surveys, and multi-country longitudinal surveys in developing countries. He is currently Research Fellow at the Institute of Statistical, Social and Economic Research at the University of Ghana.

Ola Hall is a Senior Lecturer in Social and Economic Geography at the Department of Human Geography, Lund University. His research is specialized in GIS, spatial analysis, and remote sensing.

Ellen Hillbom is Associate Professor at the Department of Economic History at Lund University. Within the broader field of African economic history, she has over the years specialized in colonial history, rural and agricultural development, and long-term changes in inequality and structural transformation. She also conducts research on current development challenges, both macroeconomic structures and local agricultural systems of production, primarily in Botswana, Tanzania, Zambia, and Malawi.

Aida Cuthbert Isinika is a Professor of Agricultural Economics at Sokoine University of Agriculture under the Institute of Continuing Education in Tanzania. She has over forty years of experience in outreach, teaching, and research on rural development. Her research work focuses on production economics and resource use efficiency. She has also published and provided consultancy services on natural resource management, gender, value chain development, and land tenure.

Magnus Jirström is a Professor of Social and Economic Geography at the Department of Human Geography, Lund University. His research is focused on rural and agricultural development in sub-Saharan Africa and Asia, which is conducted in interdisciplinary teams and in close cooperation with international research groups. In 2001 he co-founded Afrint.

Joseph T. Karugia is the Coordinator of the Regional Strategic Analysis and Knowledge Support System for Eastern and Central Africa based at the International Livestock Research Institute (ILRI) in Nairobi, Kenya. He has taught at the University of Nairobi for more than twenty years, and served as Chairman of Department of Agricultural Economics in 2002–3. Prior to joining ILRI, Joseph served as the Research Manager at the African Economic Research Consortium, with a mandate for building capacity in policy-oriented economic research in sub–Saharan Africa.

Elibariki E. Msuya is a Senior Lecturer of Economics at Sokoine University of Agriculture, School of Agricultural Economics and Business Studies. He has published book chapters as well as journal articles, specifically about unlocking smallholder potential in African agriculture.

Wapulumuka Mulwafu is Professor of Environmental History and Dean of Postgraduate Studies at Chancellor College, University of Malawi. His publications and research interests cover a wide range of issues, including the history of soil conservation, agrarian history, irrigation reform, and the socio-cultural aspects of land and water resource management. He joined the Afrint research project during the second round in 2007.

Peter Mvula is a Senior Research Fellow at the Centre for Social Research, University of Malawi. Over the past thirty years, he has done research and published in the fields of rural livelihoods, nutritional status of rural communities, food security, safety nets and social protection, health-related issues, disability, and poverty, and on disaster and relief operations using both quantitative and qualitative methods.

Willis Oluoch-Kosura is a Professor of Agricultural Economics, at the University of Nairobi. His interest is in the area of agricultural development policy, focusing on issues of rural factor and product markets, poverty dynamics, institutional analysis, and technology generation and adoption. He facilitates collaboration in teaching, research, and outreach with professional colleagues at best-practice universities and development organizations for the benefit of African universities, institutions, and agribusiness practitioners.

Martin Prowse works as an Associated Senior Lecturer in Social and Economic Geography at the Department of Human Geography, Lund University. His research focuses on agricultural commercialization, value chains, and climate change, mostly in southern and eastern Africa. He has particular expertise in the tobacco sector in Malawi and contract farming more broadly. He also has a keen interest in methodology and research methods.

Daniel B. Sarpong is Chair, Department of Agricultural Economics and Agribusiness, University of Ghana. He provides scientific analysis of the socio-economics of agricultural households and their impacts on the macro-economy.

Stephen K. Wambugu is an Associate Professor of Geography at Chuka University, Kenya. He has been a researcher, consultant, and lecturer in various academic institutions in Kenya. His areas of research interest include value chain development and actor empowerment, efficiency of agricultural markets, agricultural intensification and diversification, poverty, and food security. Stephen has worked with the other Afrint researchers since 2002.

1

Perspectives on Agriculture, Diversification, and Gender in Rural Africa: Theoretical and Methodological Issues

Agnes Andersson Djurfeldt, Fred Mawunyo Dzanku, and Aida Cuthbert Isinika

Introduction

The signing of the Maputo Declaration in 2003 marked a renewed interest in the smallholder-based model of agricultural development among a broad range of stakeholders: domestic, regional, and global – states as well as non-governmental organizations (NGOs), farmers' organizations, and researchers. This broad coalition of interests united under the banner of pro-poor agricultural growth, in the conviction that small-scale farmers if given the right policies would provide for their own food security as well as generate a marketable surplus (Mellor 1995, Lipton 2005). The eventual outcome of increasing smallholder productivity was perceived to be the gradual structural transformation of African economies, as envisaged in the 2008 *World Development Report, Agriculture for Development*, for instance (World Bank 2007).

Several challenges to the optimism of the smallholder-based growth model have arisen over the past two decades, however. While at the macro level GDP (gross domestic product) growth has been rapid and there are indeed signs of structural transformation in a number of African economies (Fuglie 2011), the inclusivity of growth processes can be questioned on several grounds. Shrinking farm sizes and increasing land size inequalities within the smallholder sector, and the emergence of middle sized farmers, are reported in studies from several African countries (Jayne et al. 2014). Nonetheless, the prospects for leaving the agricultural sector altogether are small, as shown by income data that demonstrate the persistent role of agriculture in rural livelihoods across Africa as well as the poor opportunities for diversifying into high-return activities outside agriculture (Davis et al. 2016, McCullough 2016). To such universal tendencies can be added the gender-specific aspects of exclusion based on norms and institutions that discriminate against women with respect to agricultural assets, inputs, and markets (Meinzen-Dick et al. 2014, Quisumbing et al. 2015), as well as intergenerational challenges arising from an increasingly youthful population (Losch 2012, Wiggins et al. 2015). Finally, studies attest to growing divergence in spatial terms, between well-connected, well-endowed places and more marginal areas (Andersson Djurfeldt 2013, Davis et al. 2016). The confluence of these tendencies—increasing socio-economic differentiation amid rapid economic growth (both within and outside agriculture)—could be suggestive of generally improving livelihoods, but also growing polarization within the smallholder sector over time.

To date, few studies have considered these tendencies longitudinally or in parallel—in this sense agricultural livelihoods are not situated in time or space, nor in relation to other processes such as non-farm diversification. In practice, however, rural livelihoods are characterized by sectoral as well as spatial linkages, while they are also imprinted by relations of gender and generation. Moreover, the seasonality of the agricultural calendar as well as inter-annual variations in weather and markets makes time an especially important dimension in studies of rural Africa. This book contributes through addressing the dynamics of intensification and diversification within and outside agriculture in contexts where women have much poorer access to agrarian resources than men. We use a longitudinal cross-country comparative approach to consider these linkages which have so far received limited attention in the wider literature, as well as the broader policy debates.

The book has three interrelated aims. Descriptively, the aim is to summarize and present findings from the third wave of a unique dataset – the Afrint dataset which follows smallholders across six countries from 2002 to 2013/15.

Theoretically, our aim is to provide nuance to the current dominance of structural transformation narratives of agricultural change through adding insights from gender studies as well as village-level studies of agrarian change. Placing agrarian change in relation to broader livelihood dynamics outside the farm sector and contextualizing them nationally and regionally is a necessary analytical adaptation to the unfolding empirical realities of rural Africa. We are convinced that the combination of these perspectives will enhance the research frontier in several fields and is of interest to academics within a number of disciplines.

Finally, the policy aim of the book is to provide suggestions for more inclusive policies related to rural development. Outlining the weaknesses of present policies and illustrating gendered inequalities in access to agrarian resources will provide opportunities for identifying possible alternatives to existing policy.

Theoretical Perspectives

As noted initially, the period since the early 2000s has seen the resurrection of smallholder-based approaches under the rubric of pro-poor agricultural growth. These approaches draw inspiration from the empirical example of the Asian Green Revolution and mark a distinct break with the neglect of smallholder agriculture that characterized the 'lost decades' of the 1980s and 1990s. Such models revolve around two interconnected theoretical assumptions: that small-scale farmers are efficient producers and that increased commercialization among them can encourage broad-based poverty reduction and growth. Both of these postulates are based on the dynamics of agricultural development in parts of South and South East Asia in which comprehensive rises in smallholder productivity in combination with improved markets, especially for staple crops, led to falling poverty among farmers themselves as well as landless labourers (Rosegrant and Hazell 2000, Ravallion and Datt 2002, Ravallion and Chen 2004, Djurfeldt and Jirström 2005).

At an overarching level therefore, raising smallholder productivity, enhancing commercialization, and dealing with poor producer incentives for food staples are seen as the vehicles for achieving broad-based agricultural growth and reducing poverty (Dorward et al. 2004, Jayne et al. 2006b, Diao et al. 2010, Jayne et al. 2010). Encouraging smallholder inclusion in agricultural value chains at different scales is in this respect crucial. Diversification within agriculture towards higher-value crops and a gradual movement into the non-farm sector in this way presages a gradual exit out of agriculture as the economy moves through the process of structural transformation. Where access to agricultural assets is relatively equal and initial production potential is high, rising agricultural labour productivity is more likely to emerge, enabling family members to be pulled into non-farm activities—pursuits which over time tend to be concentrated in urban centres (Hazell et al. 2007).

Challenges to the Pro-Poor Growth Model

More recently, differentiation within the smallholder sector has prompted the realization that the smallholder-based model may be inappropriate for resource-constrained households because of their limited chances of commercial engagement (Masters et al. 2013, Hazell and Rahman 2014). Here poverty reduction will not follow from commercialization or agricultural policy—rather, social policy intervention could be a better alternative.

While proponents of pro-poor agricultural growth increasingly recognize the practical limitations of the smallholder model in terms of reaching marginal households, three strands of criticism have also been levelled against the *theoretical* bedrocks of the model.

The efficiency of small-scale producers has been questioned by advocates of large-scale farming, who argue that modern technologies and procurement systems have undermined the inverse relationship between productivity and land size that underpins the notion of superior smallholder efficiency. As such, both food security and poverty reduction are better achieved through large-scale agriculture (Collier and Dercon 2014).

A second body of criticism relates to the poor fit of an Asian-inspired smallholder model to African conditions. Specifically, poor infrastructure, low initial productivity, weather-related unpredictability, and weak linkages to urban areas and industry pose considerable challenges to the possibilities for raising smallholder productivity in the same way it occurred in Asia (Ellis 2007).

A final strand of critique emanates from the scholarship on agrarian class differentiation, which questions the market optimism of the pro-poor agricultural growth model. Among these researchers, the insertion of smallholders into local and sometimes global value chains is perceived to encourage polarization of assets and incomes, as accumulation among the more well-positioned leads to the marginalization of the poor (Havnevik et al. 2007, Bernstein 2010, Bernstein and Oya 2014).

Gender and Farm Productivity

The considerable empirical evidence of gender differences in farm productivity must be added to the theoretical admonitions raised by critics of the smallholder model. The general conclusion in the literature is that systematic productivity gaps exist in favour of male-headed households, with gaps in the region of 4–40 per cent (Udry et al. 1995, Goldstein and Udry 2008, FAO 2011, Kilic et al. 2015, Slavchevska 2015). Although some of the gaps are attributable to unobservables, differences in farm input access and use account for a substantial part of the gender gaps (Doss and Morris 2001, Alene et al. 2008). The received literature also shows that gender differences in productivity have a spatial dimension, mainly based on agro-ecology (Udry 1996, Oseni et al. 2015), with the gap being substantial in low agro-productive regions where rainfall, for example, is more limiting. Whereas factors such as farm size and the presence of female family labour tends to narrow the productivity gap (Slavchevska 2015), others such as male adult labour and area cultivated to export crops tends to widen the gap (Kilic et al. 2015).

Diversifying Within and Out of Agriculture

In theoretical terms, raising smallholder productivity constitutes the basis for moving away from agriculture, as households diversify initially within and eventually out of agriculture. At the macro level the outcome of these processes is the structural transformation of the broader economy in which agriculture gradually loses its dominance in terms of employment and value added (Chenery and Syrquin 1975, Timmer 2009).

In practice, however, many rural livelihoods combine incomes from the farm and non-farm sectors, mainly through rural non-farm employment (RNFE). The rural growth linkages literature in general postulates complementarities between the farm and non-farm sectors (e.g. Haggblade et al. 1989, Delgado et al. 1994, Delgado 1998, Haggblade et al. 2007). Other studies have explored the effect of non-farm earnings or participation on farm productivity indirectly through its impact on farm input use (Savadogo et al. 1994, Hertz 2009, Mathenge et al. 2015). The empirical evidence is mixed: while some find that farm investments are increasing with non-farm earnings or participation (Lamb 2003, Ellis and Freeman 2004, Oseni and Winters 2009), others observe the opposite (Ahituv and Kimhi 2002, Kilic et al. 2009, Mathenge et al. 2015), or find no significant effect (Chikwama 2004).

The observation that there is a general decrease in farm sizes across sub-Saharan Africa (Jayne et al. 2010, Jayne et al. 2014), while at the same time diversification into RNFE is observed to be increasing (Haggblade et al. 2010, Losch et al. 2012), could lead to the conclusion that the two processes may be competitors, not counterparts. Despite these tendencies and earlier warnings of de-agrarianization and de-peasantization stemming from the literature on agrarian differentiation (Bryceson 2009), recent data on the importance of agriculture to rural livelihoods show that agriculture consistently contributes around 70 per cent of household cash incomes (Jirström et al. 2011, Davis et al. 2016, see also Chapter 2 in this volume). While engagement in the non-farm sector is high (70 per cent participation in the nine African countries covered by Davis et al. 2016), its relative contribution to rural incomes is low, pointing to the continued importance of agriculture to rural livelihoods, but also the important complementary role of non-farm diversification.

Gender and Non-farm Diversification

The general conclusion from the existing literature on gender and non-farm diversification is that participation and participation impacts are not gender blind. Whereas women tend to diversify more into non-farm self-employment than men, wage employment is more the domain of men than women (see for example Newman and Canagarajah 2000 on Ghana and Uganda). Nationally representative household surveys in Ghana for instance, have shown that participation in RNFE is higher among women than men. Gender difference in rural non-farm participation could be context-specific, however. For example, Rijkers and Costa (2012) found in their rural non-farm entrepreneurship study that whereas women were less likely to be non-farm entrepreneurs in Bangladesh, Indonesia, and Sri Lanka, no gender difference was found in Ethiopia. However, since asset endowment matters for access to high-return non-farm diversification opportunities, if asset accumulation is gendered, as indeed it is (Deere and Doss 2006, Deere 2010, Deere et al. 2013), then one would expect high-return non-farm opportunities to also be gendered, and actually this is the case (Lanjouw and Feder 2001).

Although Bagachwa and Stewart (1992) have suggested that RNFE opportunities tend to be more egalitarian than large-scale industrialization, some authors have found that the impacts of non-farm employment are not the same for women and men. For example, in Ghana, because women are more involved in non-farm self-employment, and because this type of RNFE tended to be inequality-increasing, Canagarajah et al. (2001) found that non-farm activities were inducing inequality among female-headed households rather than male-headed households. On the other hand, wage employment tended to increase inequality among men.

Even among women, non-farm participation probabilities differ between women heading their own households and women living in male-headed households. For example, being a female head of household has been found to increase the chances of non-farm labour market participation but not necessarily so for women in general (Canagarajah et al. 2001).

Dual Exclusion?

As can be inferred from the literature both on agricultural productivity as well as non-farm diversification, women face a dual exclusion based on a lack of agricultural assets, but also related to their limited access to alternative livelihood sources outside agriculture. Given the well-documented institutional bias against women with respect to ownership and control over key agricultural assets, it may be tempting to conclude that the smallholder model excludes women a priori. Studies of intra-household relations suggest, however that gendered segmentation of livelihoods may be complementary rather than conflicting, with female engagement in non-farm activities supplementing men's work in agriculture (Jackson 2007, O'Laughlin 2007). Agricultural livelihoods therefore need to be understood both in relation to gendered patterns of labour use and income generation, but also in relation to the non-farm sector. This book attempts to contribute to filling an empirical gap in this respect using data from a longitudinal dataset collected in six African countries.

Research Design

The present book constitutes a follow-up study to two earlier phases of the African Agricultural Intensification (Afrint) project. The analysis relies heavily on a quantitative dataset—collected by the Afrint group[1] in eight African countries in 2002 and 2008 (Ethiopia, Ghana, Kenya, Malawi, Mozambique, Nigeria, Tanzania, and Zambia) and again in six of these countries in 2013/15 (Ghana (2013), Kenya (2013), Malawi (2013), Mozambique (2015), Tanzania (2015), and Zambia (2013)). The data hence consist of two panel rounds (2002–8 and 2008–13/15) and three cross-sections: 2002, 2008, and 2013. Two earlier volumes have reported on Afrint I and II (Djurfeldt et al. 2005, Djurfeldt et al. 2011).

The data used in this book cover those countries for which data are available for all of the three rounds of data collection, that is Ghana, Kenya, Malawi, Mozambique, Tanzania, and Zambia. In addition, qualitative data have been collected intermittently during the second and third phase of the project but with a focus on the latter.

Quantitative Data Collection

The research design is based on a multiple-stage purposive sample, with the selection first of countries, regions second, villages third, and finally households. The original database was collected with the aim to assess the possibilities for an Asian-style Green Revolution in the context of sub-Saharan Africa (Djurfeldt et al. 2005). With this overarching objective in mind, a multistage purposive design was used to select countries and at a second stage regions that were deemed to be above average in terms of agro-ecology and

[1] <http://www.keg.lu.se/en/research/research-projects/current-research-projects/afrint>.

accessibility, but excluding the most vibrant rural economies. Within each country, variability was used as the sampling criterion for the selection of regions, such that each country sample contains regions that are both dynamic and less dynamic. The interpretation of the original sampling criteria at the country level varied somewhat—and the Malawi sample therefore contains four regions selected on the basis of crop production characteristics, while the Mozambique sample contains three rather than two regions.

Within each region, villages were again purposively selected and a random sample of the village population was taken based on household lists.[2] The sample is therefore representative at the village level. The self-identified farm manager was interviewed and data were collected for a set of household-level variables.

Data collection has been carried out at three points in time: for Afrint I, data were simultaneously collected in eight countries in early 2002, while Mozambique was added in 2005. For Afrint II, data were collected in late 2007 and early 2008 in nine countries as a follow-up to the first data collection round. For Afrint III, funding for resurveying was not available for the full dataset, moreover the funds that were secured were erratic, leading to a staggered data collection effort. Data were collected simultaneously in Ghana, Kenya, Malawi, and Zambia in early 2013, but in Tanzania in early 2015, and in Mozambique not until September of 2015. In what follows we will refer to the first round of data collection as Afrint I, the second as Afrint II, and the third as Afrint III. The first panel period (Afrint I to Afrint II) is referred to as Panel I and the second (Afrint II to Afrint III) is referred to as Panel II.

The dataset contains fifteen regions and fifty-six villages, the distribution of which are detailed in Table 1.1 and Figure 1.1.

A balanced panel design has been used to take into consideration attrition as well as changes in the village populations over time, maintaining statistical representativity between the rounds of data collection. A balanced panel design entails keeping the size and representativity of the cross-sections intact by sampling households to make up for attrition. In addition, substantial changes in the village populations between the rounds of data collection in terms of in-migration are addressed through additional sampling of in-migrants specifically. The dataset hence contains three groups of respondents: (1) panel households sampled either in two (Panel I or Panel II) or all three rounds of data collection (Afrint I, Afrint II, and Afrint III); (2) households sampled to make up for attrition (sampled in Afrint II or Afrint III); and (3) migrant households that have been added to take into consideration

[2] In the case of Tanzania, a stratified random sample was collected at the village level.

Table 1.1. Data collection regions and type, Afrint III

	Region	Region type	Number of villages	Sample size, Afrint III
Ghana	Eastern	Dynamic	4	249
	Upper East	Less dynamic	4	289
Kenya	Kakamega	Less dynamic	5	150
	Nyeri	Dynamic	5	150
Malawi	Ntchisi	Less dynamic	2	100
	Thiwi Lifidzi	Less dynamic	2	99
	Bwanje Valley	Dynamic	2	100
	Shire Highlands	Dynamic	2	103
Tanzania	Morogoro	Dynamic	5	202
	Iringa	Less dynamic	5	193
Zambia	Mkushi	Less dynamic	4	268
	Mazabuka	Dynamic	5	214
Mozambique	North	Less dynamic	4	176
	Centre	Dynamic	5	169
	South	Less dynamic	2	82

changes in the composition of the village population, either for the second or third round of data collection.

Table 1.2 provides details on sample sizes in the cross-sections and panels, and the attrition rates between the first and second and third rounds of data collection respectively.

Attrition was lower between Afrint II and III than in the previous period and also more evenly distributed between the countries.[3] The data collection efforts in Tanzania were affected by the death of the partner responsible for data collection, while in Malawi similarly, staffing issues for Afrint II explain the high rate of attrition between rounds one and two. Here the replacement of the partner between rounds two and three, like for Tanzania, has led to considerably lower attrition rates. Mozambique, finally, stands out with the highest attrition rates between Afrint II and III—the explanation is thought to lie in a combination of the high average age of the Afrint II sample and high levels of out-migration.

For the first panel, it was noted that attrition was biased on the basis of gender (more female-headed households left the sample), age (a gap of nearly twenty years was noted between those who left the sample and those who remained), education (education was higher among the household heads who left), and household size (average household size was smaller) as well as poverty (households were on average poorer) (Djurfeldt et al. 2011). While

[3] The attrition rate for Zambia between 2002 and 2008 differs from the one reported in earlier work. The reason for this is that data entry errors excluded one entire village from the survey in 2008. This village was then included again in 2013. For Panel I and II this particular village is not part of the sample, but for the Afrint I to III panel it is included.

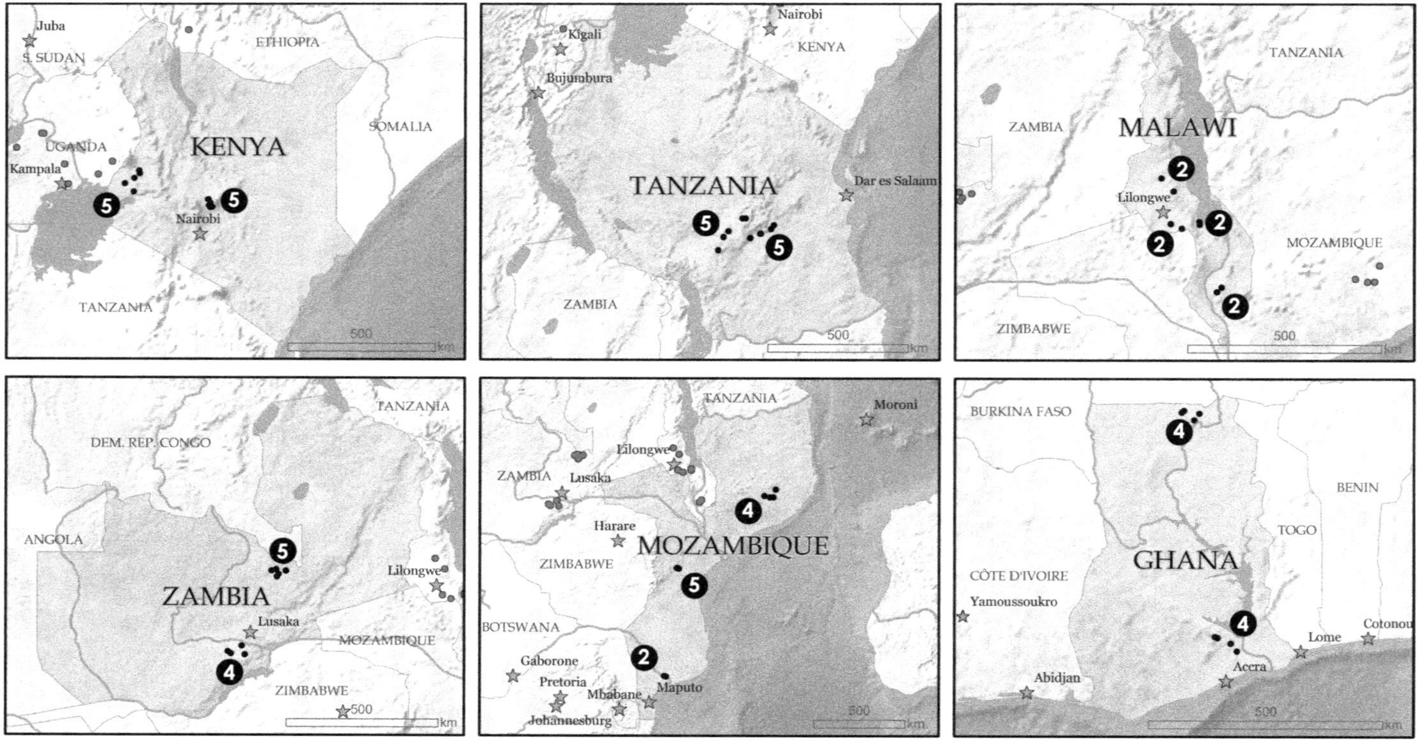

Figure 1.1 Village study sites. (Map by Maria Francisca Archila Bustos, made with Natural Earth).

Table 1.2. Number of cases in cross-section and panels, and attrition Afrint I to III

Country	Cross-sections (N)			Panels (N)			Attrition (%)		
	Afrint I	Afrint II	Afrint III	Panel I (Afrint I–II)	Panel II (Afrint II–III)	Full panel (Afrint I–III)	Afrint I–II	Afrint II–III	Afrint I–III
Ghana	417	569	538	359	491	311	14%	14%	25%
Kenya	300	300	300	266	272	251	11%	9%	16%
Malawi	400	398	402	305	346	268	24%	13%	33%
Tanzania	403	400	395	263	348	235	35%	13%	42%
Zambia	407	403	482	348	333	277	14%	18%	32%
Mozambique	398	403	427	286	295	224	28%	27%	44%
Total	2,325	2,473	2,544	1,827	2,085	1,566	21%	16%	33%

attrition was lower for the second panel, it must also be noted that these differences have largely disappeared: the share of female-headed households was the same among those who left and those who remained in the sample, as was the average age and level of education of the farm manager. In addition, as with the first panel, there was no difference in the size of cultivated area among the household types. The number of adult household members was, however, still much lower among the households that left the sample (3.1 compared with 3.4, significant at the 1 per cent level), possibly suggesting that labour was a constraining factor to production among the households that left the sample. The poverty component of attrition meanwhile has disappeared: the ability to save and the number of meals eaten in the lean season were the same for both groups, possibly suggesting that households in the second panel wave were not being pushed out of agriculture.

The Survey

The ambitions as well as the quality of the data has evolved over the project cycles: the Afrint I data focused on production and technology related to the major grain crops (rice, maize, sorghum, teff) and cassava, whereas the Afrint II survey added a more detailed section on commercialization and also collected cash income data for the first time. For the final round of data collection, cash income data were individualized, with data being collected separately for all adult household members. One challenge in capturing change over time is to deal with the fact that some variables were not gathered during the first round of data collection.

While the Afrint dataset is unique in the sense that it captures changes in rural livelihoods over time, several caveats need to be borne in mind. First, it is representative at the village level and as such is not nationally representative.

Conclusions therefore cannot be drawn about general improvements in food security, for instance. Second, we have not aimed to collect the detailed type of production data found in agricultural surveys collected by Michigan State University for instance, nor do the data provide the detail commonly found in consumption surveys (such as the Demographic Health Surveys). Rather, the dataset provides a set of broad indicators capturing changes in production, agricultural assets, food security status, commercialization, and incomes. Rural livelihood dynamics can therefore be traced through changes in these variables.

The problems associated with using the household as a sampling unit must also be acknowledged. Here, concerns can be raised based on theory (and the associated presumption of a joint utility function) as well as methodology (assuming that the household head has perfect information and as such can speak on behalf of the household). From a gender perspective the data are problematic since the food security data, for instance, do not shed light on the well-documented phenomenon of secondary poverty and food insecurity found among girls and women in households that prioritize the dietary needs of boys and men (Sen 1990). The original choice to collect data at the household level stems from budgetary as well as time constraints (neither of which have improved over time), and for reasons of comparability the unit of analysis has been retained since Afrint I.

For the final round of data collection, we have attempted to remedy the shortcomings of this approach by adding data on intra-household generation of cash income. As for Afrint III, the farm manager was asked to estimate the size of cash income for the household as a whole but individual cash income data were also collected for all adult household members within each household. Although the lack of individual data for Afrint II prevents outright comparison, differences in size and composition of cash income by sex of farm manager can be traced over time, while gender patterns of intra-household sources of livelihoods can be described for the final round of data collection at least.

The drawbacks of using cash income data to analyse livelihoods in contexts where subsistence production is widespread must also be acknowledged. Nonetheless the varied nature of production systems across regions and countries and the lack of detailed production data for roots and tubers prevent the calculation of total household income. While cash income analysis is partial it is also neutral in the sense that it does not disregard the contribution of non-grain crops to household welfare. Further admonitions also relate to the extent to which patterns of cash income generation can capture intra-household gender relations: while intra-household income data can say something about how women and men make their living, they say nothing about differences in consumption and expenditure based on sex. Moreover, even if gender-based gaps in cash income (whether within or between households) may reflect the structural discrimination of women, to what extent such shortfalls translate into differences in household welfare is not possible to assess by using income data alone.

Qualitative Data Collection

The quantitative survey has been supplemented by qualitative fieldwork carried out in two villages in Kenya (2007), four villages in Ghana (2011), three villages in Malawi (2012), and six villages in Zambia (2012 and 2016). The aim has been to shed light on intra-household aspects of livelihoods with respect to engagement in agricultural production, technology use and decision making, involvement in non-farm activities, and the control over proceeds generated from different sources of income.

The qualitative data collected in Kakamega in Kenya documented seasonality and intra-household food consumption arrangements (Andersson Djurfeldt and Wambugu 2011). Qualitative fieldwork was carried out in November 2011 in four villages in Ghana in both the Eastern and Upper East Regions. Both spouses in dual-headed households were interviewed separately, alongside women heading their own households (Andersson Djurfeldt et al. 2014). This method was replicated in November 2012 in three villages in Malawi (in Thiwi Lifidzi and Ntchisi Regions) and three villages in Zambia (in Mkushi and Mazabuka Regions), and again in another three villages in Zambia, in 2016 (Andersson Djurfeldt and Hillbom 2016). The villages in Zambia and Malawi were purposively sampled on the basis of the quantitative data as patterns in these data suggested that they had been the sites of pro-poor agricultural growth (Andersson Djurfeldt 2013).

For the intra-household interviews, respondents were selected from the quantitative sample through stratifying households by cash income per adult equivalent in each village. The households were divided into three groups on the following basis: below average, average, and above average. In each village, the intention was to select three households randomly within each category, with a total sample of nine dual-headed households (covering a total of eighteen respondents per village) in each of the six villages. The ambition was to collect data also from three households headed by single women in each income segment. Since these households were generally clustered in the below average income category, and many women had also remarried or were not possible to trace, a supplementary sample of women heading their own households was taken.

In total around 170 interviews were carried out with individual farmers across the ten villages in Ghana, Malawi, and Zambia. The individual interviews were carried out by trained research assistants—mostly masters-level students. Individual interviews at the household level were complemented with key informant interviews with the village heads (or in some cases the sub-village heads), members of the water association, lead farmers, members of credit groups and cooperatives, extension agents, and other staff from the ministries of agriculture.

To triangulate further, focus group discussions were carried out with gender-separated groups, stratified by age. The respondents in the focus groups were not part of the individual interviews. The key informant interviews and the focus group discussions were carried out by four senior researchers. The analysis in what follows is predominantly quantitative, however, with the qualitative data being used mainly to illustrate and explain the tendencies identified in the quantitative data. In particular, Chapters 3, 4, 7, and 8 draw on the qualitative fieldwork.

Who Are the Female-Headed Households?

The self-identified farm manager was interviewed during the survey. For the first round of data collection, the presumption was made that this was the same person as the household head. For the two subsequent rounds, this presumption was shown to be wrong. In the analysis that follows, the sex of the *farm manager* is used to classify households as female- or male-headed. The aspiration is to capture as broad a category of female household headship as possible, including both *de facto* and *de jure* heads of household. This objective runs the risk of merging households with quite different characteristics into a single category, however.

For Afrint III, 734 farms—or 29 per cent—were managed by women. Out of these 124 (17 per cent) had a male household head, and they were therefore *de facto* rather than *de jure* female-headed households—that is there was a male household head residing elsewhere. The use of the sex of the farm manager to classify households by gender in this sense combines households that are likely to be differently positioned with respect to the agrarian economy. Access to non-farm incomes as well as male labour may be quite different depending on whether a non-resident husband is part of the household or not. Asset data show few differences between the two household types, however, with two major exceptions. Land size is slightly larger among the *de facto* female-headed households for Afrint III: 1.7 hectares compared with 1.4 hectares for the *de jure* female-headed households (significant at the 5 per cent level), and access to male labour is also higher among the former. With respect to access to male labour, the *de facto* female-headed households replicate the patterns of male-headed households rather than the *de jure* female-headed households, with half the adult labour being male in the *de facto* female-headed households compared with 38 per cent for the *de jure* female-headed households (significant at the 1 per cent level).

Taking a closer look at the *de facto* female-headed households suggests that they are strongly concentrated to particular regions and countries with longstanding migrant labour histories. *De facto* female-headed households account for a substantial share of total female-headed households in the Centre region (54 per cent) in Mozambique, and Nyeri (30 per cent) and Kakamega (38 per cent), both in Kenya. Together these three regions account for two-thirds of the *de facto* female-headed households. While the differences between household types must be borne in mind, in general therefore the *de facto* female-headed households contribute a very minor share of the female-headed households outside these regions. At the national and regional levels, the sample sizes are too small to test the differences between *de facto* and *de jure* female-headed households statistically, but where relevant differences between the two household types will be discussed using the full sample.

Analytical Methods and Definitions of Key Terms Used in the Book

All chapters in the book are based on a common dataset, the Afrint dataset, but not all authors use the datasets for the full periods. Moreover, the depth of statistical analysis varies—some chapters rely on descriptive statistics, covering trends primarily in the cross-sectional data, whereas others use panel-level data to model causal patterns related to the overarching topics of the book: rural livelihoods and gender.

Aside from the dataset, a few unifying definitions should be noted. We use the term head of household to refer to the farm manager, hence opting for a *de facto* rather than *de jure* use of the term. The terms landholder, farm manager, and head of household are used interchangeably in the text to denote the *de facto* head of household.

A further point of classification refers to the use of the terms diversification. Here we use the term crop diversification to denote the broadening of cropping patterns—growing additional types of crops—while agricultural diversification refers to the inclusion also of livestock-based production. Income diversification describes the process of moving outside agriculture, and here we opt for the sectoral use of the terms non-farm and farm, such that farm income also includes income generated from engagement in casual farm labour on others' farms.

The geographical scope of the study is the six countries covered by the project. The discussion of findings is contextualized in relation to the broader literature on rural livelihoods in sub-Saharan Africa. At times we refer to Africa (and African conditions)—here the term is synonymous with sub-Saharan Africa.

Structure of the Book

Following this introductory chapter the book is divided into two parts: the first part covers the dataset as a whole and contains Chapters 2, 3, 4, and 5. The first three of these capture dynamics in the cross-sectional data. Chapter 2 follows up on corresponding chapters in the first two volumes (Larsson 2005, Jirström et al. 2011), describing broad trends in terms of land use, production, agricultural technology, commercialization, and non-farm diversification. Chapters 3 and 4 focus on gendered patterns of changes in asset ownership on the basis of a regional perspective and commercialization dynamics, respectively. Chapter 5 uses panel data in modelling the drivers as well as the distributional characteristics of three key processes related to structural transformation: intensification, commercial crop diversification out of staple crops, and non-farm diversification.

The second part of the book contains country studies of five of the countries covered by the quantitative dataset. Three chapters trace the role of agricultural policies for rural livelihoods over time. Chapter 6 analyses the response of the sampled farmers to input subsidies in Tanzania over time using a gender as well as geographical perspective. The linkages between agricultural policies aimed at intensification as well as crop diversification and their implications for gendered patterns of production in Malawi are discussed in Chapter 7. Finally, Chapter 8 addresses the role of price incentives versus state interventions in a comparative analysis of Zambia and Malawi.

Two chapters then focus on the linkages between agricultural productivity and non-farm incomes. Chapter 9 traces the linkages between earnings from the non-farm sector and agricultural labour productivity in Ghana from a gender perspective in combination with a regional contextualization. Similarly, Chapter 10 analyses the impact of non-farm incomes on agricultural investments and technology use for maize in Kenya.

Finally, Chapter 11 concludes the book by summarizing the empirical trends found in the dataset and reflecting on their implications for theory as well as policy.

2

African Smallholder Farmers on the Move: Farm and Non-Farm Trends for Six Sub-Saharan African Countries, 2002–15

Magnus Jirström, Maria Francisca Archila Bustos, and Sarah Alobo Loison

Introduction

Agricultural development in sub-Saharan Africa (SSA) remains high on the global development agenda. Having predominantly agriculture-based economies, the agricultural sector's role in the development and transformation of the subcontinent's economies deserves much attention in view of existing challenges regarding food insecurity and widespread poverty. Despite encouraging economic growth and transformation that has brought millions out of extreme poverty and into food security during the past fifteen years, high levels of poverty and food insecurity continue to plague Africa. With more than 40 per cent of its population, approximately 415 million, living in extreme poverty and approximately a fourth being undernourished, the SSA region faces enormous challenges (FAO 2014, ECA 2016, World Bank 2016).

Several transformational trends are currently affecting the countries of SSA and will contribute in shaping the development of their agricultural sectors in the coming decades. These trends include demographic change, rapid urbanization, and a shift in the labour force from farming to non-farm jobs. In the same way as the more recent years of successful development and change has varied between regions and countries in SSA, so, we may assume, will the impacts of the mentioned trends. Acknowledging vast regional and country differences, the broad trends mentioned can, however, be expected to bring change affecting the whole subcontinent.

The SSA population is estimated to grow from currently more than 950 million people to approximately 2.1 billion by 2050 (OECD/FAO 2016). While the rural share of the population is declining and estimated to reach 50 per cent within the next ten years, the absolute number of rural people is expected to increase for another two decades and by 2050 amount to more than 900 million (Hazell 2013, United Nations 2014). Continued rapid population growth will affect agriculture in several ways. As shown by Losch et al. (2012), between 2010 and 2025 some 330 million new labour market entrants need to be absorbed in the SSA economies. The rural share of these—195 million young people—will be looking for rural employment and if unsuccessful will leave the rural areas. In many rural settings, land fragmentation and declining farm sizes are growing phenomena and signs of growing land pressure (Jayne et al. 2014), and questions about current and future viability of minuscular farms are being raised (Hengsdijk et al. 2014).

Urbanization represents another transforming force affecting smallholder agriculture through several mechanisms. Urban consumers purchase much of the food they require, thereby potentially increasing the demand for food produced by smallholders in the rural areas. Although food imports continue to provide a very high share of overall food supply, the increasing global demand for food, fibre, and fuel from agriculture have made food imports a more expensive and risky source of supply for the growing African population. For African smallholders, this development could offer opportunities. Furthermore, urban—and increasingly also rural—people in SSA are changing diets and consuming more high-value foods such as vegetables, fruits, milk, meat, etc. These ongoing dietary shifts may provide opportunities for smallholders supplying such produce.

Urbanization in SSA is mostly characterized by the rapid growth of small and medium-sized cities. This offers opportunities for rural households to diversify their income sources and generate some of their income directly from urban and peri-urban activities without abandoning the rural areas. As will be shown later in this chapter, having access to non-farm income sources can improve smallholder household income substantially. Exiting agriculture in SSA may, as in other parts of the world, be a gradual process and there is currently strong interest in questions related to the improvement of opportunities for smallholders' income diversification into non-farm income sources (Haggblade et al. 2010, Frelat et al. 2016). Smallholders' livelihood diversification forms part of the overall structural transformation taking place in SSA. Based on a sample of more than twenty-four countries in SSA, McMillan and Harttgen (2014) estimate the share of the labour force in the agricultural sector to have declined by approximately 10 percentage points between 2000 and 2010.

A large share of the labour force in SSA agriculture is female. In a recent assessment of the contribution of women to labour in crop production across six SSA countries, Palacios-López, et al. (2016), estimate the average share at

40 per cent. Female-managed farms (FMFs) play a significant role in SSA and make up close to 30 per cent of the farms in the Afrint III sample. The final section of this chapter presents data pointing at several gender gaps contributing to the challenges facing smallholder agriculture in SSA.

This chapter uses data from the three Afrint surveys described in the methodology in Chapter 1. While panel data from the surveys are used in several chapters, this chapter reports on the data from the cross-sections. For production, the focus is set on two major SSA cereal staples—maize and rice—the production of which are of central importance for the great majority of households in the sample. As will be shown, the share of smallholder households completely depending on farming for their livelihoods remains high. At the same time, for approximately half (55 per cent) of the farm households studied, non-farm income sources provide more than half of total household cash income. While in a cross-sectional analysis of smallholder agriculture for the Afrint I and Afrint II periods, Jirström et al. (2011) could only identify flickering signs of dynamism, the presentation in the following sections will point at several important signs of dynamism, although they are clearly heterogeneous in character.

Increasing Farm Sizes: Relatively Stable Crop Portfolios

For Afrint III, the mean and median farm size, here defined as the area under cultivation, had increased substantially compared to the Afrint II survey round (Table 2.1). The Afrint III mean farm size was 2.3 ha which, compared to the Afrint II figure of 1.7 ha, represents a 36 per cent increase. The increase is significant[1] in four countries—Ghana, Tanzania, Zambia, and Mozambique—ranging between 24 per cent to 70 per cent. These countries can be categorized as relatively land-abundant in an African context (Jayne et al. 2014) and we assume that farmers in these countries have brought fallow land into production. In contrast, Kenya and Malawi, where the sampled households report small changes in farm size between the two survey rounds, belong to the growing number of land-constrained African countries. As shown in Table 2.1, the Kenyan and Malawian farm households cultivate substantially smaller farms than farmers in the other four countries.

Table 2.2 shows data for an analysis of the present household landholding size per capita, as well as a closer look at land distribution, by dividing the

[1] Statistical levels of significance are set at below 1 per cent (***), 5 per cent (**), and 10 per cent (*), respectively, throughout the book, but extreme cases have been handled differently by the authors and the treatment of extreme cases are discussed individually in each chapter.

Table 2.1. Land under cultivation, ha (mean and median)

		Afrint I	Afrint II	Afrint III	Change over time (%) Afrint I to II	Sig.	Afrint II to III	Sig.	Afrint I to III	Sig.
Ghana	Mean farm size (ha)[a]	2.41	1.99	2.84	−17	***	43	***	18	***
	Median farm size (ha)[a]	2.00	1.60	2.09						
	No. of cases	412	565	538						
Kenya	Mean farm size (ha)[a]	0.97	0.98	1.02	1		3		5	
	Median farm size (ha)[a]	0.80	0.75	0.82						
	No. of cases	300	298	300						
Malawi	Mean farm size (ha)[a]	1.24	1.18	1.21	−5		3		−2	
	Median farm size (ha)[a]	1.01	1.01	1.00						
	No. of cases	387	393	388						
Tanzania	Mean farm size (ha)[a]	1.96	1.72	2.14	−13	**	24	***	9	
	Median farm size (ha)[a]	1.60	1.40	1.62						
	No. of cases	399	392	390						
Zambia	Mean farm size (ha)[a]	3.01	2.64	3.55	−12	**	35	***	18	***
	Median farm size (ha)[a]	2.30	2.00	2.50						
	No. of cases	395	397	473						
Mozambique	Mean farm size (ha)[a]	2.02	1.33	2.25	−34	***	70	***	12	*
	Median farm size (ha)[a]	1.50	1.00	1.75						
	No. of cases	381	400	424						
Group total	Mean farm size (ha)[a]	1.98	1.69	2.30	−15	***	36	***	16	***
	Median farm size (ha)[a]	1.50	1.21	1.60						
	No. of cases	2,274	2,445	2,513						

Notes: a. Cases above 25 ha removed.

Table 2.2. Landholding size per capita

						Change over time (%)					
			Afrint I	Afrint II	Afrint III	Afrint I–II	Sig.	Afrint II–III	Sig.	Afrint I–III	Sig.
Ghana	Mean farm size (ha)[a] per capita, total sample		0.37	0.30	0.48	−18	**	57	***	29	***
	Means for household quartiles ranked by	Q1	0.09	0.09	0.12						
	farm size (ha)[a]	Q2	0.21	0.18	0.26						
	per capita by village	Q3	0.33	0.30	0.42						
		Q4	0.82	0.63	1.06						
Kenya	Mean farm size (ha)[a] per capita, total sample		0.20	0.14	0.21	−30	***	50	***	6	
	Means for household quartiles ranked by	Q1	0.05	0.03	0.07						
	farm size (ha)[a] per capita by village	Q2	0.10	0.07	0.12						
		Q3	0.17	0.13	0.19						
		Q4	0.43	0.31	0.43						
Malawi	Mean farm size (ha)[a] per capita, total sample		0.31	0.29	0.26	−5		−11	*	−16	***
	Means for household quartiles ranked by	Q1	0.12	0.12	0.09						
	farm size (ha)[a] per capita by village	Q2	0.21	0.21	0.17						
		Q3	0.31	0.30	0.25						
		Q4	0.57	0.53	0.52						
Tanzania	Mean farm size (ha)[a] per capita, total sample		0.39	0.36	0.50	−8		40	***	28	***
	Means for household quartiles ranked by	Q1	0.14	0.13	0.16						
	farm size (ha)[a] per capita by village	Q2	0.25	0.22	0.29						
		Q3	0.39	0.34	0.46						
		Q4	0.75	0.72	1.06						
Zambia	Mean farm size (ha)[a] per capita, total sample		0.43	0.38	0.50	−10		29	***	16	*

(continued)

Table 2.2. Continued

						Change over time (%)					
			Afrint I	Afrint II	Afrint III	Afrint I–II	Sig.	Afrint II–III	Sig.	Afrint I–III	Sig.
	Means for household quartiles ranked by farm size (ha)[a] per capita by village	Q1	0.15	0.12	0.15						
		Q2	0.27	0.22	0.27						
		Q3	0.43	0.38	0.43						
		Q4	0.81	0.77	1.08						
Mozambique	Mean farm size (ha)[a] per capita, total sample		0.38	0.29	0.43	−22	***	48	***	15	**
	Means for household quartiles ranked by farm size (ha)[a] per capita by village	Q1	0.12	0.09	0.12						
		Q2	0.24	0.18	0.23						
		Q3	0.36	0.27	0.39						
		Q4	0.75	0.58	0.95						
Group total	Mean farm size (ha)[a] per capita, total sample		0.35	0.30	0.41	−14	***	36	***	17	***
	Means for household quartiles ranked by farm size (ha)[a] per capita by village	Q1	0.12	0.10	0.12						
		Q2	0.22	0.19	0.23						
		Q3	0.34	0.29	0.37						
		Q4	0.70	0.60	0.88						
No. of cases[b]			2,231	2,384	2,478						

Notes: a. Cases above 25 ha removed. b. The number of cases in each quartile group varies as all households with the same per capita farm size are automatically placed in the same group in this SPSS routine.

sample into quartiles ranked by per capita farm size and by village. As can be seen, per capita access to land is very limited—in Afrint III it was 0.16 ha per capita or less for the bottom quartile (Table 2.2). For Afrint I, the corresponding figure was 0.15 ha. Under rain-fed conditions and with limited use of other land-augmenting technologies such as adequate application of inorganic and organic fertilizers, the small per capita farm size experienced by a quarter of the sample is worrying. This raises the question of whether such a limited access to farmland is sustainable, not least in light of the growth of the rural African population which is projected to continue for another two to three decades (Masters et al. 2013).

Differences between the four quartiles are large, with a range in Afrint III of six to nine times more land per capita for the highest quartile compared to the lowest one. In relation to the situation in Afrint II, the mean per capita farm size in Afrint III was 36 per cent larger, corresponding exactly to the increase in the mean farm size over the same period. Out of the six countries only the Malawi sample shows falling per capita farm sizes (−11 per cent). To our surprise, the Kenya sample does not confirm the pattern of declining per capita farm size in Kenya (Jayne et al. 2006b). Instead we find that the mean per capita access to land of the Kenyan households has risen by 50 per cent during the period. The explanation perhaps lies in the declining mean household size in the Kenyan sample from 7.0 in Afrint II to 6.1 persons per household in Afrint III.

Turning to the emergence of growing land inequalities as reported in the literature (Jayne et al. 2006a), the increasing difference in per capita farm size between the bottom and the top 25 per cent of the three Afrint samples also points in the direction of gradually growing land inequality, although the change seems moderate in the Afrint villages. For Afrint I, the per capita access to land for the bottom 25 per cent was 0.12 ha per capita and this was also the per capita size in 2013. However, the top 25 per cent expanded from a mean of 0.70 ha to 0.88 ha per capita. As a consequence, the gap between the two extreme groups has increased—in Afrint I the top quartile accessed 6.0 times as much land as the bottom quartile while in Afrint III they accessed 7.3 times as much land (see Table 2.2).

Shifting attention to land use and crop portfolios, the increase in total cultivated area between Afrint II and Afrint III corresponds to the general increase in the cultivated area of the five different categories of crops studied (Table 2.3). For Afrint III, 92 per cent of all households grew maize and the crop remained the most important crop in terms of land allocation, with an average of a little over 1 ha. Both the share of households growing maize and the maize area have increased since Afrint II.

Approximately a quarter of all households grow rice, a share which remained stable over the study period. Also in the case of rice, there has

Table 2.3. Land under cultivation (total and per crop) and share of households cultivating by type of crop

					Change over time (%)[a]					
		Afrint I	Afrint II	Afrint III	Afrint I– II	Sig.	Afrint II–III	Sig.	Afrint I–III	Sig.
Maize	Mean crop area (ha)[b]	0.80	0.81	1.08	1		32	***	34	***
	Median crop area (ha)[b]	0.50	0.60	0.80						
	Households cultivating (%)	84	86	92	2	**	6	***	8	***
	No. of cases	1,951	2,136	2,348						
Sorghum	Mean crop area (ha)[b]	0.73	0.67	0.81	–8	*	21	***	11	
	Median crop area (ha)[b]	0.50	0.50	0.50						
	Households cultivating (%)	21	20	16	–1		–4	***	–5	***
	No. of cases	487	495	409						
Rice	Mean crop area (ha)[b]	0.70	0.65	0.83	–8		29	***	18	***
	Median crop area (ha)[b]	0.50	0.40	0.60						
	Households cultivating (%)	22	23	24	1		1		2	*
	No. of cases	508	570	614						
Other food crops	Mean crop area (ha)[b]	0.56	0.51	0.79	–9		55	**	42	***
	Median crop area (ha)[b]	0.40	0.30	0.40						
	Households cultivating (%)	84	79	90	–5	***	11	***	6	***
	No. of cases	1,943	1,954	2,291						
Non-food crops	Mean crop area (ha)[b]	0.79	0.77	0.88	–2		14		12	
	Median crop area (ha)[b]	0.40	0.40	0.48						
	Households cultivating (%)	34	23	32	–11	***	9	***	–2	
	No. of cases	795	579	817						
Total	Mean crop area (ha)[b]	1.98	1.69	2.30	–15	***	36	***	16	***
	Median crop area (ha)[b]	1.50	1.21	1.60						
	Households cultivating (%)	100	100	100	n.a.		n.a.		n.a.	
	No. of cases	2,325	2,473	2,544						

Note: a. Change in percentage of households cultivating is absolute not percentage change.
b. Cases above 25 ha removed.

been a significant increase, 29 per cent, in mean hectarage. Sorghum growers also allocate more land to the crop, but the share of farmers growing the crop has dropped. Between Afrint II and Afrint III, the share dropped from 20 per cent to 16 per cent of the sample.

The single biggest increase in cropped area pertains to the category 'other food crops', which includes all food crops but maize, sorghum, and rice.[2] Land allocated to this category grew by 55 per cent between Afrint II and Afrint III, and also the proportion of households growing such crops grew from 79 per cent to 90 per cent. The category includes high-value crops including vegetables and fruits, and the increase in the cultivation of such crops can be seen as a strategy among farm households to increase the value of output per hectare. The non-food crop category, including cash crops such as coffee, tobacco, tea, and sugarcane, engages approximately a third of the sample, up from a little less than a quarter in Afrint II.

The aggregates for the six countries presented here will be explored in more detail at the country level in the subsequent sections on production and productivity for the two staples, maize and rice. We focus on these staple crops because of their strategic importance for African agriculture. The demand for food staples is growing rapidly and there is a potential for farmers in Africa to capture a large share of this growth (Hazell and Poulton 2007). Apart from contributing to growth and poverty reduction in itself, an increase in the productivity of staples can 'release land, water and labour for the production of other cash and export crops' (Hazell and Poulton 2007: 3). For African smallholders who are often net buyers of staple foods, increasing productivity in staple crop production can reduce poverty and increase food security.

Production and Area Productivity

Maize

Maize is the most widely grown staple food crop in SSA. It accounts for 30 per cent of cereal production area and about 40 per cent of cereal production (Hazell and Poulton 2007, Cairns et al. 2013). Since the early 1960s, growth in maize production has been driven mainly by area expansion, while yield growth explains only a third of the average annual growth rate of 3 per cent for the period 1961–2008 (Smale et al. 2011). However, maize yields in SSA

[2] In the Afrint III survey, cassava was included in the category 'other food crops', whereas in Afrint I and Afrint II it was not. This was due to the uncertainties and problems encountered when collecting data on cassava production in the Afrint I and II surveys. It is assumed that this change has contributed to the increasing area under 'other food crops'.

Table 2.4. Maize production (t/farm) and cultivated area (ha)

					Change over time (%)[a]					
		Afrint I	Afrint II	Afrint III	Afrint I–II	Sig.	Afrint II–III	Sig.	Afrint I–III	Sig.
Ghana	Proportion of farmers growing maize (%)[b]	43	53	71	10	***	18	***	28	***
	Mean area under maize (ha), 3-season average[c]	1.08	0.63	1.11	−42	***	76	***	3	
	Mean maize production (t/farm), 3-season average[d]	0.89	0.61	0.85	−32	***	40	***	−4	
	Median maize production (t/farm), 3-season average[d]	0.63	0.37	0.50						
	No. of cases	179	301	382						
Kenya	Proportion of farmers growing maize (%)[b]	74	59	85	−15	***	26	***	11	***
	Mean area under maize (ha), 3-season average[c]	0.52	0.61	0.51	16		−15	**	−2	
	Mean maize production (t/farm), 3-season average[d]	1.19	1.08	0.68	−9		−37	***	−43	***
	Median maize production (t/farm), 3-season average[d]	0.42	0.45	0.33						
	No. of cases	222	176	255						
Malawi	Proportion of farmers growing maize (%)[b]	94	94	93	0		−1		−1	
	Mean area under maize (ha), 3-season average[c]	0.30	0.74	0.84	144	***	13	*	176	***
	Mean maize production (t/farm), 3-season average[d]	0.70	0.83	0.89	19	***	7		27	***
	Median maize production (t/farm), 3-season average[d]	0.53	0.53	0.62						
	No. of cases	376	376	375						
Tanzania	Proportion of farmers growing maize (%)[b]	82	82	82	0		0		0	
	Mean area under maize (ha), 3-season average[c]	1.06	0.90	0.88	−15	**	−3		−17	***
	Mean maize production (t/farm), 3-season average[d]	0.98	1.18	1.29	20	**	9		31	***
	Median maize production (t/farm), 3-season average[d]	0.63	0.70	0.84						
	No. of cases	332	328	324						
Zambia	Proportion of farmers growing maize (%)[b]	77	83	88	6	**	4	*	11	***
	Mean area under maize (ha), 3-season average[c]	1.36	1.30	1.71	−4		32	***	26	***
	Mean maize production (t/farm), 3-season average[d]	1.58	2.13	3.37	34	***	58	***	113	***
	Median maize production (t/farm), 3-season average[d]	0.90	1.17	1.93						
	No. of cases	313	336	423						
Mozambique	Proportion of farmers growing maize (%)[b]	79	63	78	−15	***	15	***	0	
	Mean area under maize (ha), 3-season average[c]	0.95	0.71	1.17	−25	***	64	***	23	***

	Mean maize production (t/farm), 3-season average[d]	0.48	0.50	0.92	5		83	***	93	***
	Median maize production (t/farm), 3-season average[d]	0.35	0.35	0.50						
	No. of cases	313	255	334						
Total	Proportion of farmers growing maize (%)[b]	75	72	82	−3	**	11	***	8	***
	Mean area under maize (ha), 3-season average[c]	0.86	0.84	1.09	−3		29	***	26	***
	Mean maize production (t/farm), 3-season average[d]	0.95	1.08	1.43	13	**	32	***	50	***
	Median maize production (t/farm), 3-season average[d]	0.55	0.60	0.73						
	No. of cases	1,735	1,772	2,093						

Notes:

a. Change in proportion of households growing maize is absolute not percentage change.

b. Based on a sub-sample including maize growers who cultivated an average of at least 0.1 ha. The sample also excludes cases with a sixfold or higher yield increase between consecutive Afrint rounds.

c. Average of area under maize during the last growing season at the time of data collection and the two previous growing seasons before that one. Extreme values removed at the dataset level for each year.

d. Average of maize production during the last growing season at the time of data collection and the two previous growing seasons before that one. Extreme values removed based on average area criteria.

Table 2.5. Maize yields

		Afrint I	Afrint II	Afrint III	Change over time (%)[e]					
					Afrint I–II	Sig.	Afrint II–III	Sig.	Afrint I–III	Sig.
Ghana	Mean maize yield (t/ha),[a] 3-season average[b]	1.09	1.01	0.84	−7		−17	***	−23	***
	Median maize yield (t/ha),[a] 3-season average[b]	0.75	0.95	0.67						
	No. of cases	175	300	382						
	20% best-performing farmers' mean yield (t/ha) maize[c]	2.54	1.84	1.71	−28	**	−7		−33	***
	5% best-performing farmers' mean yield (t/ha) maize[c]	4.14	2.72	2.42	−34	*	−11		−41	**
	Average yield gap,[d] compared to 20% best-performing farmers	57	42	44	−15	***	2		−13	***
	Average yield gap,[d] compared to 5% best-performing farmers	74	59	56	−16	***	−2		−18	***
Kenya	Mean maize yield (t/ha),[a] 3-season average[b]	1.70	1.50	1.25	−12		−17	**	−26	***
	Median maize yield (t/ha),[a] 3-season average[b]	1.05	1.20	0.90						
	No. of cases	220	176	255						
	20% best-performing farmers' mean yield (t/ha) maize[c]	3.46	2.79	2.59	−19		−7		−25	**
	5% best-performing farmers' mean yield (t/ha) maize[c]	4.51	3.62	3.43	−20		−5		−24	
	Average yield gap,[d] compared to 20% best-performing farmers	52	47	52	−5		5		0	
	Average yield gap,[d] compared to 5% best-performing farmers	64	58	61	−5	*	2		−3	
Malawi	Mean maize yield (t/ha),[a] 3-season average[b]	2.22	1.17	1.16	−47	***	−1		−48	***
	Median maize yield (t/ha),[a] 3-season average[b]	1.96	0.99	0.92						
	No. of cases	374	376	372						
	20% best-performing farmers' mean yield (t/ha) maize[c]	4.11	2.19	2.33	−47	***	6		−43	***
	5% best-performing farmers' mean yield (t/ha) maize[c]	5.11	2.88	3.27	−44	***	13		−36	***
	Average yield gap,[d] compared to 20% best-performing farmers	45	44	51	−1		7	**	6	**
	Average yield gap,[d] compared to 5% best-performing farmers	56	57	62	1		5	**	6	***
Tanzania	Mean maize yield (t/ha),[a] 3-season average[b]	0.97	1.33	1.46	37	***	10	*	51	***
	Median maize yield (t/ha),[a] 3-season average[b]	0.79	1.10	1.15						
	No. of cases	332	322	323						
	20% best-performing farmers' mean yield (t/ha) maize[c]	1.85	2.59	2.83	40	***	9		53	***
	5% best-performing farmers' mean yield (t/ha) maize[c]	2.57	3.30	3.58	28	**	8		39	***
	Average yield gap,[d] compared to 20% best-performing farmers	47	49	51	2		2		4	
	Average yield gap,[d] compared to 5% best-performing farmers	61	60	61	−1		1		0	
Zambia	Mean maize yield (t/ha),[a] 3-season average[b]	1.14	1.47	1.82	29	***	24	***	59	***
	Median maize yield (t/ha),[a] 3-season average[b]	1.00	1.31	1.61						

	No. of cases	310	336	423						
	20% best-performing farmers' mean yield (t/ha) maize[c]	2.09	2.72	3.30	30	***	21	***	58	***
	5% best-performing farmers' mean yield (t/ha) maize[c]	2.72	3.39	4.28	24	***	26	***	57	***
	Average yield gap,[d] compared to 20% best-performing farmers	45	47	45	2		−2		0	
	Average yield gap,[d] compared to 5% best-performing farmers	57	57	58	0		1		0	
Mozambique	Mean maize yield (t/ha),[a] 3-season average[b]	0.53	0.83	0.89	58	***	6		68	***
	Median maize yield (t/ha),[a] 3-season average[b]	0.46	0.67	0.62						
	No. of cases	313	254	331						
	20% best-performing farmers' mean yield (t/ha) maize[c]	1.04	1.69	2.09	63	***	24	**	102	***
	5% best-performing farmers' mean yield (t/ha) maize[c]	1.45	2.39	3.31	65	***	38	**	128	***
	Average yield gap,[d] compared to 20% best-performing farmers	49	51	56	2		5		8	**
	Average yield gap,[d] compared to 5% best-performing farmers	63	64	69	1		5		6	**
Total	Mean maize yield (t/ha),[a] 3-season average[b]	1.30	1.21	1.25	−6	**	3		−4	
	Median maize yield (t/ha),[a] 3-season average[b]	0.94	1.01	0.94						
	No. of cases	1,724	1,764	2,086						
	20% best-performing farmers' mean yield (t/ha) maize[c]	2.49	2.29	2.49	−8	*	9	**	0	
	5% best-performing farmers' mean yield (t/ha) maize[c]	3.36	3.02	3.39	−10		12	**	1	
	Average yield gap,[d] compared to 20% best-performing farmers	48	47	49	−2		3	**	1	
	Average yield gap,[d] compared to 5% best-performing farmers	61	59	61	−3	**	2	**	−1	

Notes:

a. Based on a sub-sample including maize growers who cultivated an average of at least 0.1 ha. The sample also excludes cases with a sixfold or higher yield increase between consecutive Afrint rounds.

b. Average of maize yield during the last growing season at the time of data collection and the two previous growing seasons before that one.

c. Based on village aggregates.

d. The yield gap is calculated as the difference between the household's 3-season average yield and the average of the best-performing farmers in the village.

e. Change in average yield gap is absolute not percentage change.

remain far behind the global average, and at around 1.8 t/ha only reach a third of the global mean maize yield of around 5.5 t/ha (CIMMYT 2016, USDA 2016). The mentioned average maize yield of 1.8 t/ha stems from Food and Agricultural Organization (FAO) data. Serious criticism of the quality of the national agricultural statistics in SSA on which the FAO data build, has been levelled (Carletto et al. 2014). The farm household survey data on which our figures build also point at significant discrepancies in comparison with the FAO data. When making comparison with national-level data this should be kept in mind.

The three-season average production per farm in our sample of six countries increased substantially from 1.08 t per farm in Afrint II to 1.43 t in Afrint III (Table 2.4). The overall growth in production of 32 per cent is largely the result of an increase in the maize area which grew by 29 per cent during the period. The level of median production volume, 0.60 t in Afrint II and 0.73 t in Afrint III, indicates that, in spite of substantial growth, there is not much, if anything, to market for many households considering that maize is an important food for most households. Except in Zambia where the maize area is larger, the median total production is in the range of 0.33 t to 0.84 t.

The overall mean yield of maize for the entire sample of maize growers of 1.25 t/ha for Afrint III does not indicate any significant change in yields since Afrint I (1.30 t/ha) or Afrint II (1.21 t/ha) (Table 2.5). Variation in production and productivity among the six countries is substantial. In Ghana, Mozambique, and Zambia the mean maize area, mean production, and share of households growing maize increased substantially between Afrint II and Afrint III (Table 2.4).

In Zambia, the mean maize yield increased by 24 per cent between Afrint II and III, a positive change which is close to the national increase of 22 per cent for the same period (FAO 2016). For Mozambique, change during the period was not statistically significant, while for Ghana area expansion outpaced production increase to the extent that the three-year average yield of 0.86 t/ha for the Afrint III period represents a decrease of 17 per cent compared to the Afrint II level. Also, national figures for Ghana show that the average maize yields are among the lowest in the world and remain much lower than the average for SSA (Ragasa et al. 2014). As shown by Dzanku and Sarpong (2014), regional differences in the Ghana sample are substantial, with farmers in the four villages situated in the Eastern Region achieving yields approximately twice as high as those in the Upper East Region in the northern and more arid parts of the country.

In Kenya, the proportion of households planting maize in the Afrint III sample had increased since Afrint II. However, the mean maize area decreased by 15 per cent and the production fell even more (−37 per cent) and, thus, the mean maize yield declined by 17 per cent from 1.5 t/ha to 1.25 t/ha in our study regions between Afrint II and III. At a national level, yields have remained steady, at around 1.5 t/ha over the period of our study, setting Kenya, together with Tanzania, apart from neighbouring Eastern African countries such as Ethiopia, Uganda, and Malawi where maize yields have risen since 2003 (Aylward et al. 2015).

In Tanzania and Malawi, change between Afrint II and Afrint III has been limited and is not statistically significant. For Tanzania, the mean yield of 1.46 t/ha in Afrint III is close to the national level of 1.38 t/ha for 2012–14 and, as in our sample, the change in mean yields since 2006–8 has been marginal (FAO 2016, FAOSTAT). In the case of Malawi, the sampled households reported maize yields only about half the level of the national three-year mean yields of approximately 2.2 t/ha for the Afrint III period. A possible explanation for this difference may relate to the sampling criteria for the Malawi sample, where maize was the major crop only in some of the villages.

As previously indicated, average maize yields in SSA are low by international comparison. This is true also in comparison to most other tropical regions where maize is grown under rain-fed conditions. But yield variability also sets SSA apart. According to Byerlee and Heisey (1997), yield variability is almost always higher in SSA countries than in other developing countries with similar mean yields. While climatic factors such as major regional droughts are important explanations, price variability and government policies as well as interventions can be assumed to play a role as well. Yield variability does, however, not only occur between countries and regions. As shown in Table 2.5, differences in maize yields at the village level are significant. The six-country three-season average for Afrint III shows that the yield gap at the village level is substantial. The best-performing 20 per cent of farmers in each village achieve, on average, the double maize yield compared to each individual farmer in the village. When comparing each farmer's yield with the 5 per cent best-performing farmers' yield, the gap widens and reaches around 60 per cent. In a subsequent section, we will return to the yield gap discussion in more detail. First, we turn to the second most important cereal among the Afrint farm households, namely rice.

Rice

The importance of rice is growing rapidly in SSA, and demand is growing more rapidly than in any other continent (Nasrin et al. 2015). Although production increased at an annual rate of 3.3 per cent over the period 1970–2009, the gap between consumption and production widened as consumption grew by an annual rate of 4 per cent and thus outpaced production. In 2009, 37 per cent

Table 2.6. Rice production (t/farm) and cultivated area (ha)

					Change over time (%)[a]					
		Afrint I	Afrint II	Afrint III	Afrint I–II	Sig.	Afrint II–III	Sig.	Afrint I–III	Sig.
Ghana	Proportion of farmers growing rice (%)[b]	40	29	33	−11	***	4		−7	**
	Mean area under rice (ha), 3-season average[c]	0.66	0.59	0.87	−10		46	***	32	**
	Mean rice production (t/farm), 3-season average[d]	0.51	0.30	0.88	−41	***	194	***	72	***
	Median rice production (t/farm), 3-season average[d]	0.35	0.17	0.35						
	No. of cases	165	163	177						
Malawi	Proportion of farmers growing rice (%)[b]	31	25	26	−6	**	1		−5	*
	Mean area under rice (ha), 3-season average[c]	0.44	0.48	0.54	10		13		24	**
	Mean rice production (t/farm), 3-season average[d]	0.76	1.07	0.91	40	***	−15		19	
	Median rice production (t/farm), 3-season average[d]	0.66	0.97	0.67						
	No. of cases	93	74	77						
Tanzania	Proportion of farmers growing rice (%)[b]	44	43	42	−1		−1		−2	
	Mean area under rice (ha), 3-season average[c]	0.99	0.97	1.13	−2		16	**	13	*
	Mean rice production (t/farm), 3-season average[d]	1.51	1.89	2.65	25	**	41	***	75	***
	Median rice production (t/farm), 3-season average[d]	1.17	1.37	2.09						
	No. of cases	176	171	170						
Zambia	Proportion of farmers growing rice (%)[b]	1	0	0	−1		0		−1	
	Mean area under rice (ha), 3-season average[c]	0.32	n.a.	n.a.	n.a.		n.a.		n.a.	
	Mean rice production (t/farm), 3-season average[d]	0.23	n.a.	n.a.	n.a.		n.a.		n.a.	
	Median rice production (t/farm), 3-season average[d]	0.23	n.a.	n.a.						
	No. of cases	4	0	0						
Mozambique	Proportion of farmers growing rice (%)[b]	11	15	24	4		9	***	13	***
	Mean area under rice (ha), 3-season average[c]	0.41	0.29	0.59	−29	***	103	***	45	***
	Mean rice production (t/farm), 3-season average[d]	0.14	0.23	0.22	63	**	−5		55	**
	Median rice production (t/farm), 3-season average[d]	0.10	0.15	0.16						
	No. of cases	44	60	103						

Group total	Proportion of farmers growing rice (%)[b]	21	19	21	−2	2		0	
	Mean area under rice (ha), 3-season average[c]	0.71	0.67	0.85	−5	26	***	19	***
	Mean rice production (t/farm), 3-season average[d]	0.89	0.99	1.33	11	34	***	49	***
	Median rice production (t/farm), 3-season average[d]	0.56	0.50	0.53					
	No. of cases	482	468	527					

Notes: a. Change in proportion of households growing rice is absolute not percentage change.

b. Based on a sub-sample including rice growers who cultivated an average of at least 0.1 ha. The sample also excludes cases with a sixfold or higher yield increase between consecutive Afrint rounds.

c. Average of area under rice during the last growing season at the time of data collection and the two previous growing seasons before that one, extreme values removed at the dataset level for each year.

d. Average of rice production during the last growing season at the time of data collection and the two previous growing seasons before that one, extreme values removed based on average area criteria.

of consumption was imported and SSA has become a major importer of rice with almost a third, 11.8 million tons, of rice traded globally (Nasrin et al. 2015). Rice is now the second largest source of food in SSA and since the turn of the century has gradually come close to maize as a provider of food calories in the subcontinent. The growing popularity and consumption of rice can be traced to a combination of population growth, urbanization and urban growth, and changing diets.

Approximately a fifth of the Afrint sample cultivate rice, the share of which has been stable since Afrint I (Table 2.6). Out of the six countries, two—Kenya and Zambia—lacked households growing rice during the Afrint III period. Among the remaining four, Ghana and Tanzania together accounted for more than 70 per cent of the 527 rice-farming households during the Afrint III period. In the two countries, there has been a significant increase in the mean cultivated rice area, mean farm production, and mean yields over the three-season periods of Afrint II and Afrint III (Tables 2.6 and 2.7). The recorded growth tallies well with the national-level data for the two periods (FAO 2016, FAOSTAT). Data on milled rice consumption per capita for Ghana and Tanzania show a very rapid increase between 1982 and 2012 (Nasrin et al. 2015), and it can be assumed that the strong growth in demand for rice contributes to the substantial increases in production and productivity among the surveyed farms in Ghana and Tanzania.

In the Mozambique sample, the share of households growing rice increased from 15 per cent to 24 per cent between Afrint II and Afrint III. The area under rice approximately doubled from 0.29 ha to 0.59 ha per farm but production remained at around 0.22 t/farm. Consequently, the mean three-season rice yield fell from 0.90 t/ha in the Afrint II period to 0.52 t/ha in the Afrint III period. FAO data indicate a similar situation at the national level, with a dramatic increase in the national rice hectarage accompanied by a sharp fall in the average national rice yield during the period (FAO 2016, FAOSTAT). In Malawi, finally, the fifth of the surveyed households cultivating rice saw their yields declining by, on average, 19 per cent, from 2.25 t/ha to 1.81 t/ha, during the period. National data point at stable rice production and productivity for the period (FAO 2016, FAOSTAT).

Yield Gaps

One out of several ways to conceptualize the term yield gap is to view it as the differences between the average crop yield (area productivity) achieved by farmers and what is achievable using more optimal cultivars, inputs, and other management practices. The level of productivity in staple crop production across countries in SSA is generally far below that of developing countries in other regions. While this situation has been observed for many years, more recently attention has also been drawn to the large differences in crop yields achieved within the subcontinent. Yield gaps in sub-Saharan Africa are large, among the largest globally, and particularly so in the case of maize (Mueller and Binder 2015). Tittonell and Giller (2013) estimate current yield gaps for major food crops in Africa, and conclude that yield gaps for almost all crops in all regions remain wide and are likely to increase if soil degradation continues.

One reason for the surging focus on yield gaps relates to the growing awareness that yield growth in already cultivated areas of the world needs to increase at a fast rate in order for global food production to keep pace with demand. Demand is expected to grow substantially in the coming years due to a combination of population and income growth. The realization that yield growth has begun to stagnate in many areas in both developed and developing countries explains the growing research interest in agricultural intensification and, as a part of this complex, in the analysis of existing yield gaps (Beddow et al. 2014, FAO and DWFI 2015). Narrowing yield gaps in SSA might also contribute to the post-2015 global Sustainable Development Goals of ending hunger and achieving food security by 2030.

The Afrint data on yields for the two staple crops, maize and rice, concur with the observations about large yield gaps in SSA. In order to get an indication of what is achievable under local conditions, we have defined the attainable yield as the mean yield of the 20 per cent best-performing farmers per crop and village (outliers excluded). The average difference between that level and each individual farmer's yield we label the yield gap. Unlike the more common use of the term, we do not refer to the difference between the actual and the agronomic yield potential of the crop (FAO and DWFI 2015). Rather we use the concept yield gap to capture the difference between what is already the attainable yield under local conditions and realities—bio-physical as well as socio-economic—and what is the actual yield for each farmer in the village. This, we think, gives a clear indication of the potential for improvement of yields at the local level.

The aggregates based on village means of the top 20 per cent farms with the highest yields for the three staple cereals are presented in Tables 2.5 and 2.7. As shown, the gaps are substantial in Afrint III—49 per cent for maize and 46 per cent for rice. The percentages indicate the discrepancy between the village attainable yield and the yield of the farmers in the village. For maize, the data for Afrint III show that the top 20 per cent of farms attained yields on average two times higher than the individual farmers' yield. Comparing the situation over the three different time periods, we observe no marked change. The gaps

Table 2.7. Rice yields (ton/ha)

		Afrint I	Afrint II	Afrint III	Change over time (%)[e]					
					Afrint I–II	Sig.	Afrint II–III	Sig.	Afrint I–III	Sig.
Ghana	Mean rice yield (t/ha),[a] 3-season average[b]	0.98	0.48	0.88	−51	***	81	***	−11	
	Median rice yield (t/ha),[a] 3-season average[b]	0.76	0.35	0.62						
	No. of cases	164	163	177						
	20% best-performing farmers' mean yield (t/ha) rice[c]	2.03	0.90	1.61	−56	***	79	***	−20	***
	5% best-performing farmers' mean yield (t/ha) rice[c]	2.64	1.15	1.97	−56	**	71	**	−26	
	Average yield gap,[d] compared to 20% best-performing farmers	50	46	44	−4		−2		−6	
	Average yield gap,[d] compared to 5% best-performing farmers	61	60	57	−1		−3		−4	
Malawi	Mean rice yield (t/ha),[a] 3-season average[b]	1.79	2.25	1.81	26	**	−19	**	1	
	Median rice yield (t/ha),[a] 3-season average[b]	1.57	2.26	1.56						
	No. of cases	93	74	77						
	20% best-performing farmers' mean yield (t/ha) rice[c]	3.28	3.53	3.62	8	***	2		10	***
	5% best-performing farmers' mean yield (t/ha) rice[c]	4.19	3.63	4.86	−13		34		16	
	Average yield gap,[d] compared to 20% best-performing farmers	48	49	55	1		6		7	
	Average yield gap,[d] compared to 5% best-performing farmers	63	59	65	−3		6		3	
Tanzania	Mean rice yield (t/ha),[a] 3-season average[b]	1.52	1.95	2.24	28	***	15	**	47	***
	Median rice yield (t/ha),[a] 3-season average[b]	1.45	1.61	2.09						
	No. of cases	176	171	170						
	20% best-performing farmers' mean yield (t/ha) rice[c]	2.58	3.31	3.55	28	***	7		37	***
	5% best-performing farmers' mean yield (t/ha) rice[c]	3.08	4.00	4.61	30		15		50	***
	Average yield gap,[d] compared to 20% best-performing farmers	42	40	37	−3		−3		−5	
	Average yield gap,[d] compared to 5% best-performing farmers	54	52	52	−2		0		−2	

Zambia	Mean rice yield (t/ha),[a] 3-season average[b]	0.47	n.a.	n.a.	n.a.		n.a.		n.a.	
	Median rice yield (t/ha),[a] 3-season average[b]	0.40	n.a.	n.a.						
	No. of cases	4	0	0						
	20% best-performing farmers' mean yield (t/ha) rice[c]	0.77	n.a.	n.a.	n.a.		n.a.		n.a.	
	5% best-performing farmers' mean yield (t/ha) rice[c]	0.77	n.a.	n.a.	n.a.		n.a.		n.a.	
	Average yield gap,[d] compared to 20% best-performing farmers	34	n.a.	n.a.	n.a.		n.a.		n.a.	
	Average yield gap,[d] compared to 5% best-performing farmers	34	n.a.	n.a.	n.a.		n.a.		n.a.	
Mozambique	Mean rice yield (t/ha),[a] 3-season average[b]	0.39	0.90	0.52	134	***	−42	***	36	
	Median rice yield (t/ha),[a] 3-season average[b]	0.25	0.63	0.33						
	No. of cases	43	60	103						
	20% best-performing farmers' mean yield (t/ha) rice[c]	0.76	1.90	1.20	150	***	−37	*	58	***
	5% best-performing farmers' mean yield (t/ha) rice[c]	0.72	2.61	1.59	262	***	−39		121	**
	Average yield gap,[d] compared to 20% best-performing farmers	49	49	56	0		6		7	
	Average yield gap,[d] compared to 5% best-performing farmers	57	60	64	3		4		7	
Total	Mean rice yield (t/ha),[a] 3-season average[b]	1.28	1.35	1.38	5		2		8	
	Median rice yield (t/ha),[a] 3-season average[b]	1.06	0.97	1.05						
	No. of cases	480	466	525						
	20% best-performing farmers' mean yield (t/ha) rice[c]	2.30	2.37	2.42	3	***	2		5	***
	5% best-performing farmers' mean yield (t/ha) rice[c]	2.60	2.78	2.87	7		3		10	
	Average yield gap,[d] compared to 20% best-performing farmers	47	45	46	−2		1		−1	
	Average yield gap,[d] compared to 5% best-performing farmers	58	57	58	−1		1		0	

Notes: a. Based on a sub-sample including rice growers who cultivated an average of at least 0.1 ha. The sample also excludes cases with a sixfold or higher yield increase between consecutive Afrint rounds.

b. Average of rice yield during the last growing season at the time of data collection and the two previous growing seasons before that one.

c. Based on village aggregates.

d. The yield gap is calculated as the difference between the household's 3-season average yield and the average of the best-performing farmers in the village.

e. Change in average yield gap is absolute not percentage change.

are within the range of 47 per cent to 49 per cent for maize and 45 per cent to 47 per cent for rice.[3]

One argument in favour of a stronger focus on yield gaps relates to poverty reduction, and suggests that by closing or reducing yield gaps, poor farmers trapped in low-productivity farming would benefit. As shown by Dzanku et al. (2015), drawing on Afrint panel data for twenty-one regions in eight SSA countries, poverty gaps do increase as yield gaps grow. Whether investment made to close such gaps would be the most efficient way to reduce poverty among African smallholders remains, however, uncertain. The same question also seems relevant when addressing another and related argument, namely that, by closing yield gaps, food security would improve for smallholders, particularly among subsistence farmers. As discussed in a comprehensive analysis of more than 13,000 small farms in seventeen SSA countries,[4] a strategy to improve market access and to increase off-farm opportunities would be more efficient in improving food security than a focus on agricultural production and closing yield gaps (Frelat et al. 2016). Among the several possible factors explaining yield gaps, the use of productivity-increasing technologies is often emphasized. Among these, the use of improved seed varieties and fertilizers receive a lot of attention. In the following section, the focus is set on these technologies.

Technology Adoption

Sustainable intensification has become a lead concept in agricultural development generally, and specifically in the discussion of agricultural development in SSA. Although area expansion will remain important for increased food production for several decades to come, it is broadly recognized that there is an urgent need for strategies targeting sustainable intensification of already cultivated land. This recognition is partly the result of two current and simultaneous trends, namely the growing land pressures in many countries and regions within countries and the unsustainability of current low-productivity agriculture which is based on nutrient mining, causing different forms of land degradation and stagnating yields.

[3] If comparing each farmer's yield with the 5 per cent top-yielding farms the yield gaps increase to between 59 per cent to 61 per cent for maize and 57 per cent to 58 per cent for rice. Because the Afrint village samples contain between thirty and fifty farms, a 5 per cent subsample will contain one to three farms only and the statistical error could be high, so we prefer to use the 20 per cent level when analysing the village sample yield gaps; however, we provide the 5 per cent values here to maintain comparability with previous publications.

[4] The Afrint dataset forms part of data used in the study.

It is in this context that new technology and technology adoption becomes particularly important in an SSA perspective. Globally, the introduction of new productivity-increasing technologies has been central for agricultural development during the past half century. Large parts of tropical and subtropical Asia and Latin America benefited tremendously from intensification processes popularly referred to as the Green Revolution (Djurfeldt and Jirström 2005). During the same period, development and diffusion of new technologies in Africa generally lagged behind, although there are several well-documented examples of productivity spurts sharing several features with those of the Green Revolution (Holmén 2005). Adoption rates are therefore currently lower in SSA than in countries in Asia and Latin America; nevertheless, many African smallholders are familiar with, for example, new seed and fertilizer technologies (Sheahan and Barrett 2014).

Focusing in this section on the adoption of seed technology and the use of inorganic fertilizer, we show that adoption rates of improved maize seeds and for fertilizer were quite high during the period studied (Table 2.8).[5] In the Afrint III period, 50 per cent of maize farmers used improved seeds and 55 per cent applied inorganic fertilizer. The use of improved rice varieties was 20 per cent in Afrint III, a halving since Afrint II. Fertilizer application had, however, become more common with 30 per cent of farmers applying it in Afrint III compared to 19 per cent in Afrint II. For the two crops, the share of households investing in fertilizer use has grown and the usage in the Afrint III period is quite high.

Possibly related, it can also be noted that since the mid-2000s several African countries subsidize fertilizer at high rates. Despite what may seem a somewhat confusing stand by the World Bank as to whether governments should or should not subsidize fertilizer programmes (Mellor 2014), all but Mozambique of the Afrint countries have such programmes. In Mozambique, where the use of fertilizer is very low and most farmers are not familiar with the technology, there is not a general programme but pilot tests have been conducted and an expansion of test areas is taking place (Simtowe 2015).

Technical advances such as the use of improved seed or employment of improved nutrient management can trigger and drive increases in surplus farm production that can be marketed. According to Wiggins et al. (2011), technical advances can be seen as one of two important drivers of the commercialization of small-scale farming, the other being higher prices and better access to markets. We now turn our attention to these issues.

[5] Some caution is warranted in the interpretation of the figures presented in Table 2.8. Farmers reporting on the use of improved seed sometimes refer to recirculated seeds. On the other hand, farmers sometimes refer to improved seed as traditional because they have used it for several years and consequently label it traditional technology.

Table 2.8. Seed and fertilizer use (share of famers using)

						Change over time					
			Afrint I	Afrint II	Afrint III	Afrint I–II	Sig.	Afrint II–III	Sig.	Afrint I–III	Sig.
Maize	All farmers	Proportion using improved seed (%)	45	53	50	8	***	−3	**	5	***
		Proportion using fertilizer (%)	41	41	55	0		14	***	14	***
	20% best-performing farmers[a]	Proportion using improved seed (%)	47	62	54	15	***	−8	**	7	**
		Proportion using fertilizer (%)	50	46	57	−4		11	***	7	***
	5% best-performing farmers[a]	Proportion using improved seed (%)	53	61	55	8		−7		1	
		Proportion using fertilizer (%)	48	50	59	2		9		11	
Rice	All farmers	Proportion using improved seed (%)	18	40	20	21	***	−20	***	2	
		Proportion using fertilizer (%)	17	19	30	1		11	***	12	***
	20% best-performing farmers[a]	Proportion using improved seed (%)	18	45	24	28	***	−21	***	7	
		Proportion using fertilizer (%)	19	22	31	3		8		12	**
	5% best-performing farmers[a]	Proportion using improved seed (%)	15	42	23	27	***	−19	*	8	
		Proportion using fertilizer (%)	24	23	33	−1		10		8	

Note: a. Based on mean yield, 3-season average.

Commercialization and Market Integration

Although far from all observers consider commercialization as an inherently positive process,[6] the most common policy stand taken by governments and international development organizations like the World Bank is to promote the commercialization of the smallholder sector. This standpoint is based on the notion that commercialization generally will benefit the farmers involved, and also contribute to economic growth and improved livelihood opportunities outside the farm economy (World Bank 2007). Considering Africa's population growth rate and the increasing numbers of urban residents depending on the market for their food needs, it would be hard not to appreciate the importance of agricultural commercialization.

Given the heterogeneity in terms of access to land, capital, labour, and skills found in many smallholder communities, commercialization does not unfold evenly across farms. This should be kept in mind when analysing data on commercialization. When right after the harvest, for example, a low-income farm household sells a high share of its maize in spite of knowing that it will soon have to buy maize but at a higher price, its level of commercialization may indeed have increased, but such a distress sale tells us that it may not represent a positive change for that household.

With this reflection as a cautionary backdrop, data on crop producers' participation in the market as well as the share of produce marketed are presented in Table 2.9. We also show the absolute amount of staple cereals sold. The great majority (84 per cent) of all households sold some of their crops in Afrint III. The share has grown over the period studied. As expected, non-food crop producers are very market-oriented, with 87 per cent participating in the market. Also, the share of commercialized growers of food crops other than maize, sorghum, and rice is large, with 70 per cent of these producing for the market. By generating immediate income through the year for cash-strapped households, the sale of crops like vegetables, beans, and potatoes is important. Looking at its overall importance as a source of cash income for the entire sample (not only growers), the category made up on average 20 per cent of total cash income in the Afrint III sample, and sale of these types of food crops was thereby the second most important source of cash income.

The cereal staples, for which we have richer data, continue to be important sources of cash income for many households. Since Afrint II, the proportion of growers selling cereal crops has increased. Approximately half of all maize growers, which includes 92 per cent of all households, sell maize. For rice

[6] Stockbridge (2006) and Wiggins et al. (2011) offer reviews of agricultural commercialization in Africa which include discussions of different and sometimes competing perspectives on the social and economic impacts of commercialization.

Table 2.9. Share of farmers selling and amount marketed by type of crop

		Afrint I	Afrint II	Afrint III	Change over time (%)[a]					
					Afrint I–II	Sig.	Afrint II–III	Sig.	Afrint I–III	Sig.
Maize	Proportion of growers who sell the crop (%)	35	45	49	11	***	4	***	15	***
	Average amount sold, sellers (t)[b]	0.63	0.98	1.44	55	***	47	***	127	***
	Median amount sold, sellers (t)[b]	0.30	0.36	0.51						
	Average proportion of total production sold, all farmers (%)	16	39	23	23	***	−16	***	7	***
Sorghum	Proportion of growers who sell the crop (%)	20	4	10	−16	***	6	***	−10	***
	Average amount sold, sellers (t)[b]	0.21	0.10	0.18	−53	***	86	*	−13	
	Median amount sold, sellers (t)[b]	0.11	0.06	0.10						
	Average proportion of total production sold, all farmers (%)	6	2	3	−5	***	2	**	−3	***
Rice	Proportion of growers who sell the crop (%)	61	51	58	−9	***	7	**	−3	
	Average amount sold, sellers (t)[b]	0.59	0.69	1.32	18		89	***	123	***
	Median amount sold, sellers (t)[b]	0.30	0.42	0.60						
	Average proportion of total production sold, all farmers (%)	26	25	30	−1		5	***	4	**
Other food crops[c]	Proportion of growers who sell the crop (%)	65	73	70	8	***	−2	*	6	***
Non-food crops[c]	Proportion of growers who sell the crop (%)	–	96	87	n.a		−9	***	n.a	
Any type of crop[c]	Proportion of growers who sell the crop (%)	71	80	84	8	***	4	***	12	***

Notes: a. Change in proportion of growers who sell the crop and average proportion of total production sold is absolute not percentage change.
b. Extreme values removed at the dataset level.
c. Data on amount sold was not available.

growers the share is higher (58 per cent), while only a tenth of sorghum growers market the crop. Part of the explanation for the low share of marketed sorghum may be the popularity of the crop for beer production, which opens up possibilities for on-farm processing and sale at the local level.

Between Afrint II and III, the average amounts of maize and rice sold increased substantially and, for the Afrint III sample, 1.44 t of maize and 1.32 t of rice were the average amounts sold per farm (all sellers considered). Comparing means and medians, the skewed distribution of marketed output implies that, for the great majority, the amounts marketed are modest, and in most cases limited to a few bags. As noted earlier, this implies that, for many, the sale of these staples is followed by later purchases, eventually turning many sellers into net buyers. The average proportion of total production sold was very low for sorghum, presumably for the reason previously mentioned. For maize and rice we can note that some 23 per cent of all maize produced and 30 per cent of rice were marketed. In the study by Frelat et al. (2016) referred to earlier, covering more than 13,000 small farms in SSA, the share of food crops sold was 20 per cent, a level that our results tally well with.

An indication of the relative importance of cereals staples as a cash earner is that in both Afrint II and Afrint III staple food sales constituted the single most important source of cash income, with an average share of 24 per cent in Afrint II and 26 per cent in Afrint III. For the six countries, only in Kenya were staple crops not the most or second most important cash income source. Taken together, the sale of crops on average made up more than half (54 per cent) of all cash income in the Afrint III sample. Adding the share from the sale of animal products (9 per cent), agricultural production represented 63 per cent of total cash income among the Afrint households. When including agricultural work on others' farms as a part of farm income, the total share of farm income in relation to total cash income reaches 70 per cent, ranging from 49 per cent in Mozambique to 85 per cent in Zambia. This share comes close to estimates by Haggblade et al. (2007) and Jayne et al. (2006a). The relative importance of farm income sources has not changed since Afrint II, a sobering reminder of the heavy reliance on agriculture in the livelihoods of African smallholders.

Non-Farm Linkages and Diversification

African smallholders in rural areas[7] are often farming in regions where economic sectors other than agriculture are significant. Estimating the non-farm share of the rural economies across the developing world,

[7] Long neglected, African urban farmers are increasing in number and importance. In this volume, however, the focus is on smallholders in rural areas.

Haggblade et al. (2010) show that 35 per cent of rural household income in Africa is derived from non-farm sources. Moreover, in their review, Barrett et al. (2001) indicate an even larger figure of between 40 per cent and 45 per cent for rural Africa. For Asia and Latin America their estimate is approximately 50 per cent. Landless and near-landless households are particularly dependent on the rural non-farm economy, but non-farm income sources are also important for smallholders who have access to more land.

Much hope is presently being pinned on the development and growth of the rural non-farm economy in SSA. By providing pathways out of poverty, such a development seems imperative in order to meet challenges such as pervasive rural poverty, decreasing farm sizes, and the rapid growth in the number of young rural people entering the labour market (Losch et al. 2012). In its analysis of the role of agriculture in contributing to poverty reduction and inter-sectoral growth in SSA, the *World Development Report 2008* is explicit when pointing out that agriculture cannot be expected to drive the growth and transformation of the rural economies in all regions where smallholders operate (World Bank 2007). In areas with low potential for agricultural development, other development strategies must, according to the report, be sought, including the provision of safety nets by governments. Not even in the more medium- and high-potential areas (in terms of agro-ecology and market access) can all smallholders be expected to farm their way out of poverty. The *World Development Report 2008* does not, however, enter into the challenging questions of how to finance the proposed safety nets.

For resource-poor smallholders, the diversification of livelihoods is often a necessity caused by economic distress, rather than a strategy to seek attractive income opportunities. Moreover, even for the non-farm sector, substantial entry barriers limit access to high-return rural non-farm income to relatively better-off households, while the poor are mainly confined to low-return non-farm activities. In a review of rural livelihood diversification in SSA, Alobo Loison (2015: 1134) concludes that while the benefits of the diversification into non-farm income sources mainly favour the better-off, diversification 'still provides a safety net for the rural poor and sometimes offers a means for upward mobility'.

The Afrint II and Afrint III surveys provide data on different types of cash incomes sources. In Table 2.10 these are grouped into farm and non-farm sources. Following a classification of income sources according to economic sectors, agricultural labour is here included in the overall category 'farm income'. On average, non-farm income sources account for 30 per cent of total household cash income.[8] There is variation between the country

[8] Cash income only forms part of total household income for households retaining part of their agricultural output for their own consumption, payment for hired labour, seeds, etc. If this part of

aggregates, with the proportion of non-farm income (NFI) in the Afrint III period being the lowest in Zambia (15 per cent) and the highest in Mozambique (51 per cent). In the case of Zambia, it is the dramatic increase in food staple production and sales that have reduced the overall share of NFI from 29 per cent to 15 per cent between Afrint II and Afrint III. In Mozambique, the NFI share was stable over the period and its greater importance is due to the domination of subsistence farming.

While on average NFI makes up an important share of total cash income, still many households report not having any NFI at all (Table 2.10). As much as 44 per cent of the Afrint III sample did not report any NFI. Ranging on average between 50 per cent and 60 per cent, the share of households which do generate NFI has been stable during the period. In Afrint III, two countries diverged from this level. In Ghana, 75 per cent of households engaged in non-farm activities, and in Zambia only 39 per cent reported having NFI. In Zambia, the proportion of households participating in the non-farm sector dropped from 56 per cent in Afrint II to 39 per cent in Afrint III, a change which may reflect the boom in maize production. In Ghana, the share of households engaging in micro-business increased significantly (16 per cent), contributing to the growing importance of that source of income, which increased from 8 per cent to 16 per cent of total household cash income between the Afrint II and Afrint III periods.

Production, Commercialization, and Income Diversification by Gender and Income Group

Following the two themes of this volume, this section of the chapter presents findings from a cross-tabulation of indicators for production, productivity, commercialization, and diversification comparing FMFs and male-managed farms (MMFs) as well as four income groups. The analysis depicts income patterns and income diversification among the sampled farm households on the basis of these comparisons. More detailed analysis of gender differences of production, commercialization, and income diversification at the country and regional levels follows in Chapters 3 and 4.

Gender Differences

As noted in Chapter 1, FMFs constitute 29 per cent of the Afrint III sample. The survey data show that landholdings held by women were only approximately

the production was to be valued at the price of marketed output and added to the amount of cash income, the overall share of farm income would be markedly higher.

Table 2.10. Non-farm income shares

					Change over time					
		Afrint I	Afrint II	Afrint III	Afrint I–II	Sig.	Afrint II–III	Sig.	Afrint I–III	Sig.
Ghana	NFI as a share of total income (%)	–	28	39	n.a.		11	***	n.a.	
	Share of households having NFI (%)	49	54	75	5		22	***	26	***
Kenya	NFI as a share of total income (%)	–	18	26	n.a.		8	***	n.a.	
	Share of households having NFI (%)	70	32	56	−38	***	24	***	−14	***
Malawi	NFI as a share of total income (%)	–	32	23	n.a.		−8	***	n.a.	
	Share of households having NFI (%)	67	60	52	−7	*	−8	**	−15	***
Tanzania	NFI as a share of total income (%)	–	33	26	n.a.		−7	***	n.a.	
	Share of households having NFI (%)	71	50	50	−21	***	0		−21	***
Zambia	NFI as a share of total income (%)	–	29	15	n.a.		−14	***	n.a.	
	Share of households having NFI (%)	52	56	39	4		−17	***	−13	***
Mozambique	NFI as a share of total income (%)	–	48	51	n.a.		3		n.a.	
	Share of households having NFI (%)	54	47	60	−6	*	13	***	6	*
Total	NFI as a share of total income (%)	–	31	30	n.a.		−1		n.a.	
	Share of households having NFI(%)	60	51	56	−9	***	5	***	−4	***

half the size (56 per cent) of those held by men (Table 2.11). For maize, the mean cultivated area of FMFs of 0.86 ha constitutes 69 per cent of that of MMFs. This difference explains most of the gender difference in maize production with male-headed households producing 1.64 t per farm compared with 0.86 t for their female counterparts. In terms of area productivity, the gap is not big. In Afrint III, FMFs achieved 90 per cent of the maize yield of MMFs, a gap which was the same in Afrint I.

Gender-based differences in the adoption rate of seed and fertilizer technology in maize cultivation were not very large in Afrint III (Table 2.11). The share of FMFs using improved seeds and inorganic fertilizers was 45 per cent and 49 per cent respectively. For male-headed households, the corresponding proportion of users was 51 per cent and 56 per cent. However, lacking detailed data on use, for example, amounts and types of fertilizers, we cannot further explore potential differences in technology application.

Turning to the indicators of commercialization, the difference is larger with only 40 per cent of female farm managers selling maize while 53 per cent of the male respondents sold some of their maize in the Afrint III period. Furthermore, FMFs tend to sell a smaller share of their maize—17 per cent compared to 25 per cent for the MMFs. In comparison to the Afrint II period, the share of maize sold declined substantially to 32 per cent and 41 per cent for FMFs and MMFs respectively, but the gap between the two remained at the same level. The gap in the level of commercialization can perhaps be related to the much smaller total production on FMFs and the fact that, in many households, maize is a popular and important food staple. In this context, we can also note that while male landholders on average set 45 per cent of the farm area to maize, FMFs used 55 per cent of their land for maize in the Afrint III period. The pattern for maize is similar for the crop category of other food crops. In the case of non-food cash crops, the difference is smaller with 81 per cent of the FMFs marketing their produce compared to 88 per cent of the MMFs.

Gender differences in income and income diversification are substantial, with FMFs generating annual cash incomes/per adult equivalent to only approximately 70 per cent of those of MMFs (see also Chapter 4). For MMFs, on average 72 per cent of total cash income derives from farm sources while for FMFs the share is 63 per cent. Being more dependent on NFIs for their total income, FMFs also depend more on agricultural wages. For Afrint III, work on others' land generates 9 per cent of total income for FMFs while only 4 per cent for MMFs.

Diversification by Income Group

By dividing the farm households into four groups according to their cash income sources—farm only, non-farm only, both, or none—several important

Table 2.11. Production, commercialization, and diversification by sex of farm manager

	Afrint I				Afrint II				Afrint III			
	Male	Female	Diff.	Sig.	Male	Female	Diff.	Sig.	Male	Female	Diff.	Sig.
No. of households	1,609	712			1,811	661			1,803	734		
Proportion of sample	69	31			73	27			71	29		
Mean farm size (ha)	2.21	1.45	0.77	***	1.90	1.13	0.77	***	2.63	1.49	1.13	***
Mean area under maize (ha), 3-season average[a]	0.96	0.67	0.28	***	0.93	0.61	0.32	***	1.19	0.82	0.37	***
Mean maize production (t/farm), 3-season average[b]	1.07	0.72	0.35	***	1.21	0.72	0.50	***	1.64	0.86	0.78	***
Mean maize yield (t/ha),[c] 3-season average[d]	1.34	1.20	0.14	**	1.23	1.17	0.07		1.29	1.15	0.14	***
Proportion using improved maize seed (%)	46	43	3		55	49	6	**	51	45	6	***
Proportion using fertilizer on maize (%)	39	45	−6	**	41	39	3		56	49	7	***
Proportion of maize growers selling maize (%)	39	26	13	***	48	38	11	***	53	40	13	***
Proportion of total maize production sold (%)	18	11	7	***	41	32	9	***	25	17	8	***
Proportion of growers of other food crops selling (%)	68	56	12	***	75	67	8	***	74	60	14	***
Proportion of growers of non-food crops selling (%)	–	–			96	96	0		88	81	7	**
Proportion selling any crop (%)	76	62	14	***	83	72	11	***	88	73	15	***
Sale of staple crops as a proportion of total cash income (%)	–	–			26	19	7	***	28	21	7	***
Agricultural wage labour as a proportion of total cash income (%)	–	–			5	9	−4	***	5	9	−4	***
NFI as a proportion of total cash income (%)	–	–			29	39	−10	***	28	37	−10	***
Mean household income per adult equivalent[e], 2010 PPP-adjusted USD[f]	–	–			281	297	−16		317	226	91	***

Notes: a. Average of area under maize during the last growing season at the time of data collection and the two previous growing seasons before that one.
b. Average of maize production during the last growing season at the time of data collection and the two previous growing seasons before that one.
c. Based on a sub-sample including maize growers who cultivated an average of at least 0.1 ha. The sample also excludes cases with a sixfold or higher yield increase between consecutive Afrint rounds.
d. Average of maize yield during the last growing season at the time of data collection and the two previous growing seasons before that one.
e. The use of adult equivalents takes the age composition of the household into account, by assigning adult household members (16–60) a value of 1, whereas children (0–15 years of age) are given value of 0.50 and elderly household members (61 years and above) are given a value of 0.75, when calculating the number of adult equivalents per household.
f. Afrint II incomes are adjusted from 2008 values; Afrint III incomes are adjusted from 2013 values for Ghana, Kenya, Malawi, and Zambia, and 2015 values for Tanzania and Mozambique. Extreme values removed at the village level.

observations indicating substantial differences in terms of production, productivity, and commercialization and total cash income can be made (Table 2.12). The great majority of households belong either to the group relying completely on agriculture for cash income (40 per cent) or the group combining farm and NFI sources (49 per cent). Only 7 per cent of households report that they depend completely on non-farm sources of income. The remaining 5 per cent report no cash income at all.

Starting with the biggest group, households combining farm and NFIs, the average total household income of this group is approximately double that of the groups relying completely on either farm or NFI. This finding is broadly in line with previous research pointing to the benefits of supplementing farm incomes with NFIs among African smallholders (Alobo Loison 2015). Compared to the group relying completely on NFI to generate cash, the group combining income sources has larger mean farm size (2.44 ha compared to 1.47 ha), attains higher mean maize yields (1.26 t/ha compared to 0.85 t/ha), and more commonly uses improved maize seeds and fertilizer on maize.

The 40 per cent of farm households completely dependent on agriculture for cash income do not show statistically significant differences from the ones combining farm and NFI sources in terms of production, productivity, and commercialization. The only significant difference is in the proportion using improved maize where the farm-only group has a significantly higher proportion of households using improved maize seeds than any of the other groups.

Likewise, the group of 116 households reporting not to have any cash income shows no significant difference from the group that depends on NFI in several characteristics, including farm size, maize yield, and a less common usage of improved seeds and fertilizers. Both groups, however, show significantly lower results for all of these factors as compared to the groups of households whose cash incomes are specialized in agriculture and those having both agricultural and NFIs. Conclusions about this group of households reporting to have no cash income must, however, be drawn with care. Several of the households reported that they sold or intended to sell some of their crop produce, and thus the true number of households that belong to this group can be questioned (see note in Table 2.12). Nonetheless, we note that this group of households appears to be very resource weak.

Concluding this section, we emphasize the finding that the large group of households that are completely dependent on farm activities for cash income, on average have low cash incomes in spite of good access to land, a relatively high share of households using improved seeds and inorganic fertilizer for maize, and are comparatively commercialized. This prompts a question which decision-makers and policymakers need to consider: would the farm households in this group be able to farm their way out of poverty if they had better

Table 2.12. Production, commercialization, and diversification by income group

	Afrint I				Afrint II				Afrint III			
	Only farm income	Only non-farm income	Both income sources	No income	Only farm income	Only non-farm income	Both income sources	No income	Only farm income	Only non-farm income	Both income sources	No income
No. of households	831	305	1,082	107	1,096	252	1,006	119	1,006	182	1,240	116
Proportion of sample	36	13	47	5	44	10	41	5	40	7	49	5
Mean farm size (ha)	2.30	1.31	1.98	1.39	1.72	1.07	1.86	1.27	2.37	1.47	2.44	1.38
Mean area under maize (ha), 3-season average[a]	0.94	0.70	0.88	0.70	0.80	0.67	0.91	1.01	1.11	0.92	1.10	0.87
Mean maize production (t/farm), 3-season average[b]	0.96	0.48	1.13	0.58	1.03	0.59	1.25	1.18	1.68	0.64	1.37	0.62
Mean maize yield (t/ha),[c] 3-season average[d]	1.25	0.94	1.48	0.87	1.22	0.91	1.31	1.01	1.35	0.85	1.26	0.76
Proportion using improved maize seed (%)	43	24	54	38	55	24	61	30	57	22	49	28
Proportion using fertilizer on maize (%)	38	26	48	28	39	16	51	21	61	20	56	24
Proportion of maize growers selling maize (%)[e]	43	5	40	13	51	13	51	16	56	5	54	6
Proportion of total maize production sold (%)[e]	20	2	17	6	43	5	45	17	28	1	24	4
Proportion of growers of other food crops selling (%)[e]	76	10	73	21	78	22	79	39	78	18	74	35
Proportion of growers of non-food crops selling (%)[e]	–	–	–	–	96	100	96	95	92	45	88	35

Proportion selling any crop (%)[e]	85	13	82	19	89	25	86	45	93	25	89	35
Sale of staple crops as a proportion of total cash income (%)	–	–	–	–	36	0	18	0	39	0	19	0
Agricultural wage labour as a proportion of total cash income (%)	–	–	–	–	10	0	3	0	10	0	4	0
NFI as a proportion of total cash income (%)	–	–	–	–	0	100	49	0	0	100	45	0
Mean household income per adult equivalent,[f] 2010 PPP-adjusted USD[g]	–	–	–	–	243	141	371	0	213	179	373	0

Notes: a. Average of area under maize during the last growing season at the time of data collection and the two previous growing seasons before that one.
b. Average of maize production during the last growing season at the time of data collection and the two previous growing seasons before that one.
c. Based on a sub-sample including maize growers who cultivated an average of at least 0.1 ha. The sample also excludes cases with a sixfold or higher yield increase between consecutive Afrint rounds.
d. Average of maize yield during the last growing season at the time of data collection and the two previous growing seasons before that one.
e. The data are based on households' answers as to whether they had sold or intended to sell any maize following the most recent crop year. The classification of household in the four categories shown in the table is based on respondents' answers about incomes from different income sources during the past year. This explains why a number of households belonging to the income categories 'only non-farm income' and 'no income' are included as sellers of crops. The number of cases in Afrint III for maize was 8 and 6 respectively, for other food crops 28 and 34, for non-food crops 14 and 8, and for any crop sold 44 and 39. The number of cases in Afrint II for maize was 28 and 17 respectively, for other food crops 31 and 36, for non-food crops 10 and 20, and for any crop sold 62 and 54.
f. The use of adult equivalents takes the age composition of the household into account, by assigning adult household members (16–60) a value of 1, whereas children (0–15 years of age) are given value of 0.50 and elderly household members (61 years and above) are given a value of 0.75, when calculating the number of adult equivalents per household.
g. Afrint II incomes are adjusted from 2008 values; Afrint III incomes are adjusted from 2013 values for Ghana, Kenya, Malawi, and Zambia, and 2015 values for Tanzania and Mozambique. Extreme values removed at the village level.

access to affordable technologies, better access to markets, and better credit and insurance facilities?

Conclusion

Since Afrint II the approximately 2,500 farm households studied have increased their agricultural production substantially. On average, the production of maize, sorghum, and rice increased by more than 30 per cent. This was made possible by a roughly corresponding expansion of the area cultivated with these crops, but also other food crops such as vegetables, roots and tubers, and fruits. Thus, the mean farm area cultivated increased by 36 per cent and reached 2.30 ha in 2013. FMFs, which make up approximately 30 per cent of all farms in the sample, are significantly smaller than male-held landholdings, however. Cultivating only nearly half as large areas, and, in the case of maize, attaining about 90 per cent of the yield per hectare of male-headed households, female-headed households produce substantially less farm output.

Productivity measured in terms of output per area did not change significantly on average. There is variation between countries but broadly for the whole sample and for the whole period between Afrint I and Afrint III, changes in production depend generally on corresponding changes in the area cultivated. While yields for the main cereal staples remain low, the gap between the top performers in each of the fifty-six villages and the great majority is substantial, with the top 20 per cent of farms producing approximately twice as much per area. This indicates that higher yields are attainable under local conditions and realities—biophysical as well as socio-economic—and thus there is a clear potential for raising local yields.

The general increase in crop production resulted in significant increases in the maize and rice volumes marketed. The overall proportion of farm households who marketed some share of their different crops harvested increased since the Afrint II survey, reaching 84 per cent in Afrint III. Taken together, the sale of crops, on average, makes up more than half of total household cash income and, adding to this other agricultural income sources, the proportion of farm income to total income reaches 70 per cent.

To the extent that this description of increases in production and the levels of commercialization signals positive change since Afrint II, a broader view of smallholders' livelihoods shows that farm households relying entirely on income from agricultural sources are faring much worse than those able to combine farm and non-farm cash income sources. Making up some 40 per cent of the total sample, smallholder farms lacking NFI generate only about half as much cash income as those farm households also able to find employment and opportunities in the non-farm economy. In spite of cultivating, on average, more than 2.3 ha and participating in the different crop commodity markets, these households do not, under current conditions, seem to be able to farm their way into the income levels of households with more diversified income sources.

Returning to this chapter's introduction, and the several major transformatory trends affecting the predominantly agrarian economies of SSA, the picture emerging from our broad description of development in the smallholder sector of six SSA countries clearly points to the need for reinforced support to the sector. At the same time, we have shown that by supplementing farm income with NFI, half of the households studied are able to increase their total income significantly. This points to the opportunities for supportive policies and initiatives targeting the development of the rural non-farm economy to contribute to improved conditions for African smallholders.

3

Assets, Gender, and Rural Livelihoods

Agnes Andersson Djurfeldt

Introduction

A growing body of literature details the importance of assets to poverty reduction as well as enhanced welfare and empowerment. A sub-field of this scholarship recognizes the constraints to women's control and ownership of assets (Quisumbing and Maluccio 2003, Deere and Doss 2006) and the impact as well as importance of assets in agricultural development interventions focusing on gender (Quisumbing et al. 2015, Johnson et al. 2016).

Given that this is a relatively new field, the empirical evidence on asset ownership and control is limited in general and the evolution of assets over time even more so. This chapter therefore has a largely descriptive aim of presenting trends in asset control and ownership over time in the six countries covered by the study. The chapter should be seen as complementary to Chapters 2 and 4, with the focus being physical assets of importance to rural livelihoods. This choice stems from the limitations of the dataset and the mainly quantitative methodology of the study, which does not permit the analysis of more intangible assets such as access to kinship networks or associational capital. All data will be discussed on the basis of gender, where asset ownership, access, and control will be compared on the basis of the sex of farm manager. As noted in Chapter 1, this includes both *de facto* and *de jure* female-headed households. Where relevant the two groups of female-headed households will be compared.

While the limitations of this approach were recognized in Chapter 1, unfortunately the data do not permit an analysis of intra-household ownership and control over assets. This would clearly have been preferable as numerous studies attest to the importance of cultural norms in restricting women's ownership as well as control over assets *within* households (Meinzen-Dick et al. 2014, Johnson et al. 2016).

A second caveat relates to the limited possibilities for addressing aspects of asset control (rather than just ownership)—with the exception of land, the dataset does not contain information on asset control. Bearing these shortcomings in mind, we are able to trace changes in asset ownership among the smallholders in the study over time to assess whether livelihoods have improved or stagnated, and whether these changes have affected male- and female-managed farms (MMFs and FMFs) differently.

While assets enable raising incomes, the accumulation of assets over time also indicates improvements in livelihoods. The analysis therefore reflects on both the changing potential to generate income (whether in cash or kind) and withstand shocks, but also the outcome of this potential over time. After a brief review of key concepts and previous literature, the chapter is structured according to this logic, starting with a discussion of the key agricultural assets of land, labour, livestock, and water resources, followed by an analysis of non-agricultural assets that can be used to generate incomes. Finally, outcomes of livelihood strategies are considered through analysing data on non-productive consumer durables and the ability to save.

Gender, Assets, and Rural Livelihoods

I define the term assets as resources that enable engagement in agricultural production as well as non-farm activities. This is broadly in keeping with the notion of capitals found in the Sustainable Livelihoods Framework (Chambers and Conway 1992, Scoones 1998, Scoones 2009). More recently, contributions have gendered the livelihoods framework, with the work of Meinzen-Dick et al. (2014) being especially inspiring in this regard (see Meinzen-Dick et al. 2014: 100).

Broadly speaking the livelihood literature that has inspired as well as built on the Sustainable Livelihoods Framework (Chambers and Conway 1992, Scoones 1998, Bebbington 1999, Rakodi 1999, Scoones 2009, Meinzen-Dick et al. 2014), identifies a set of 'capitals' through which households and individuals are able to source their livelihoods. The composition of livelihoods therefore depends on the availability and use of these different types of capitals, something that in turn is also related to the natural resource base, the institutional environment, and access to public goods. Aspects of culture and religious norms related to gender roles may also influence who has access to and control over particular capitals.

Several variants of the original framework exist, each recognizing slightly different types of capital. Scoones (1998) in an early working paper identifies four types of capital: natural, economic/financial, human, and social, but notes that this is not an exhaustive list. Bebbington's (1999) work distinguishes between natural and produced capital, adds cultural capital, but excludes economic capital. Meinzen-Dick et al. (2014) describe six key capitals: natural, physical, human, financial, social, and political capital. Other recent studies distinguish between agricultural and non-agricultural assets, and household durables and agricultural durables (Quisumbing et al. 2015).

For the purpose of this chapter, one key distinction that cuts across the various types of capital is the difference between agricultural and non-agricultural assets. Figure 3.1 presents a schematic representation of how the division between agricultural and non-agricultural assets can be reconciled with the capitals commonly used as a point of departure in livelihood studies.

The analysis in what follows focuses on natural, physical, and human capital and to a lesser extent financial capital. The dataset, moreover, has a strong slant towards agricultural rather than non-agricultural assets.

Deere (2010) identifies a number of ways in which assets are linked to poverty: productive assets can be used to generate incomes, rent, interest, and profits. In the case of agriculture, production can also be used for own consumption, while assets provide use value. Assets can be used as collateral for credit and also constitute a safeguard against shocks and possibilities for storing value. In the context of strong seasonality and widespread vulnerability, asset ownership can therefore smooth the impact of crises, while enabling households to take hold of opportunities that arise when economies grow (Deere and Doss 2006: 1).

In feminist economic theory, the ownership over assets is linked to the fall-back position of the individual within the household. While households are sites of cooperation as well as conflict, access and control over resources increases the individual's bargaining power, since the person in control of assets can threaten to leave the household, hence diminishing the resource base of the household as a whole (Quisumbing 2010). Asset ownership is therefore connected to women's possibilities for influencing household

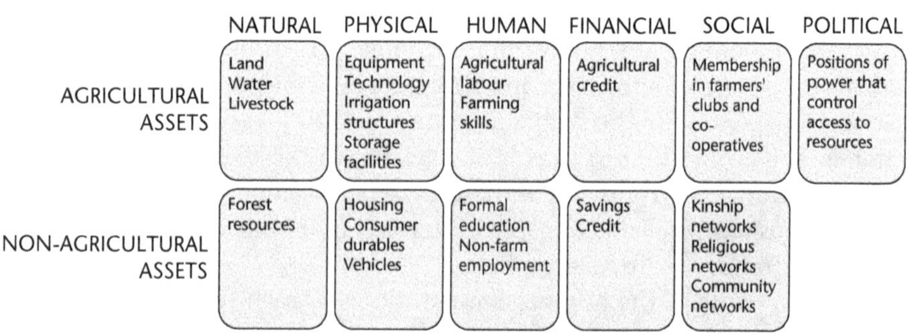

Figure 3.1 Capitals and assets.

Notes: a. Examples of assets are included in the figure—these are not exhaustive.

decision making and enhancing their opportunities both within and outside the household. Moreover, increased female control over assets is empirically connected to improvements in several development metrics, both for women individually as well as their families (Quisumbing and Maluccio 2003). While increasing women's access to and control over assets holds substantial potential for poverty reduction, the literature demonstrates that men generally hold more assets as well as higher-value assets than women, a situation attributed to gender biases in the norms that regulate ownership of assets (Deere and Doss 2006).

Agricultural Assets

As suggested by Figure 3.1, agricultural assets cut across several capitals, although the importance of each particular capital to local livelihoods varies depending on broader institutional and environmental contexts. As noted initially, data are not available for all assets, which limits the analysis in terms of the amount and type of assets that can be discussed. Referring back to Figure 3.1, data on storage facilities are limited, as are data on social capital. Moreover, the inability to discuss aspects of asset control clearly restricts the depth of analysis, a fact that is especially problematic in contexts characterized by 'bundles of rights' to basic agrarian resources, such as land and water. Keeping such reservations in mind, the following sections discuss the key agricultural assets of land, labour, water, livestock, and technology use.

Land

Several gender asset gaps are noted in the literature, with most attention having been paid so far to gender gaps in land access and control over land. In general, private ownership of land is limited among African smallholders, with land rights being based largely on community land allocations, where holders have user rights to land tied to their membership of a particular tribe, clan, or lineage.

LANDHOLDERS, LAND RIGHTS, AND SECURITY OF TENURE

As with many other parts of the world, women generally access land through marriage, with female land rights being secondary in the sense that their husbands hold the right to cultivate the land. Communal land rights in many parts of sub-Saharan Africa are based on patrilineal systems where land transfers exclude women. Marital regimes across most African countries follow separation of property, meaning that property is held individually and upon dissolution of the union assets revert to each party. In practice this means that following a divorce, women in many countries do not have access to land brought by their husbands into the marriage. Widow's rights to land are also generally weak, especially for women who are not indigenous to the particular community (Lastarria-Cornhiel et al. 2014).

There are exceptions to these general patterns, however, with the matrilineal land systems found in parts of Malawi, Zambia, and Mozambique being perhaps the most obvious. Even in these contexts, however, the rights to land and control over the proceeds from that land vary considerably. Peters (2010) notes that post-marital residence is a crucial aspect of tenure, with matrilineal descent and uxorilocal settlement (in the wife's village) being connected to the strongest female rights to land. Another exception is provided by parts of Ghana, where the inheritance of land by daughters means that both spouses bring land into marriage and continue to use and control particular plots within the household also during marriage. In the Tanzanian study sites, clan-based land rights provide generally secure rights for women, even if land systems are patrilineal. Finally, several of the study countries, most obviously Kenya and parts of Zambia, have examples of private freehold tenure, although the rights for women within these systems are not necessarily stronger than those found in areas under community-based tenure.

Available agricultural census data depart from the FAO definition of a landholder, which is defined as the person who 'makes major decisions regarding resource use and exercises management control'. The landholder therefore can be, but is not always the owner of the land, as the landholder may be using land under communal tenure, or renting or borrowing land. As suggested by Lastarria-Cornhiel et al. (2014: 126), the assumption is made that the household controls the holding, which 'does not take into account the cultivation of separate plots by men and women, and the possibility that men and women may have different rights to land, even within the same household'.

The FAO definition of landholder largely overlaps with the sex of farm manager that we use to classify farms as male-managed or female-managed. The data reported from the Afrint survey are therefore largely comparable with the FAO statistics. Lastarria-Cornhiel et al. (2014) present data from the FAO gender and land rights database for four of the countries covered by the Afrint study: Malawi, Mozambique, Tanzania, and Zambia.

On average, for Afrint III, 29 per cent of the sample consisted of FMFs, with country-level shares varying little over time. Based on the variation in female land rights noted in the literature and the data compiled by Lastarria-Cornhiel et al. (2014), the expectation would be that Malawi, Zambia, and Mozambique would have relatively high shares of female land-holders. Mozambique indeed stands out,with 43 per cent FMFs, which diverts strongly from the FAO data where 23 per cent of landholdings were held by women. The FAO data are quite old (from 1999/2000), however, and the post-conflict movement and loss of men in the civil war may explain the higher figure for Afrint III. The share of female farm managers for Malawi (31 per cent) corresponds almost exactly with that reported by the FAO (32 per cent), whereas the Afrint data for Zambia show a substantially higher share (23 per cent) than the figure of 15 per cent reported by the FAO for 1990. Tanzania, with a share of 24 per cent FMFs, corresponds quite closely to the FAO data (20 per cent). In Kenya, 36 per cent of the farms in the sample were managed by women, but this may be related to migration patterns in the particular sample sites, where men have traditionally left to work outside the villages. Finally, the lowest share of female farm managers is found in Ghana (21 per cent). The admonitions provided by Lastarria-Cornhiel et al. 2014 regarding female land rights *within* male-headed households are in order for the final case, since several of the sample sites are part of land use systems in which particular plots are individually managed within the household.

While the share of FMFs provides information on the land cultivated by female farm managers and their family members, security of tenure may be different for female landholders and land use may also be differentiated by gender. The property rights literature discusses the 'bundles of rights' that pertain to land and distinguishes between use rights, rights to the proceeds generated from the land, decision-making rights over land management, and the right to transfer land to other people (Meinzen-Dick and Mwangi 2009, Lastarria-Cornhiel et al. 2014).

The dataset contains information on some of these rights: the issue of control has been addressed in all survey rounds, with respondents in general stating that they have full control over the land that they cultivated, not needing to obtain permission for cultivation or changes in crops and land use. For Afrint I, 8 per cent of the sampled respondents stated that they needed permission, whereas the remainder claimed to have full control. By the time of Afrint II, the share of the sample who were required to get permission had increased somewhat to 12 per cent, but dropped slightly to 10 per cent for Afrint III. Statistically significant differences based on gender of farm manager emerged only for the last data collection round, with 11 per cent of the male farm managers stating that they needed permission compared with 9 per cent among the female farm managers (significant at the 10 per cent level). For the latest data collection round, female farm managers had slightly stronger rights to use the land without interference from persons outside the household. Formal titles to land or certificates of registration were held by 23 per cent of the respondents during Afrint II, with a slight increase by the final data collection round, when the corresponding figure was 29 per cent for male farm managers, compared with 25 per cent for female farm managers (significant at the 5 per cent level). To the extent that formalization of tenure has increased, it has therefore biased MMFs.

A tenth of the households surveyed during the third phase stated that they had lost land that they considered they had the right to cultivate, but there were no statistically significant differences based on the sex of the farm manager. Somewhat surprisingly—given that land is commonly accessed through husbands or male kin—there were no statistically significant differences in land losses or control over land between *de facto* and *de jure* female-headed households.

While female farm managers are in the minority across all the sampled countries, their user rights and control over land appear to be largely equal to those of male farm managers. The gender asset gap may in this respect be larger *within* some male-headed households than between the two farm types, as suggested by qualitative work from Malawi which shows the limited right exercised by women to returns from land both in terms of food and income in male-headed households.

SIZE OF LANDHOLDINGS

Gendered patterns of landholding size can shed light on the gender asset gap in the study areas as well as the relative possibilities for women and men to engage in agriculturally based livelihoods.

Numerous studies attest to the regional nature of agriculturally based livelihoods, with production patterns being based both on physical aspects such as soil fertility and rainfall, as well as institutional characteristics tied to tenure systems and economic factors such as access to markets and infrastructure (Wiggins 2000). With respect to gender specifically, the point is often made that cultural and social norms that influence women's access to and control over natural resources are highly contextual, sometimes varying from one village to another. Because of the small sample sizes at the village level (and the even smaller number of female farm managers sampled at this level), the data will be treated at the regional rather than the village level, however.

As noted in Chapter 1, the data were collected in fifty-six villages across fifteen regions. While data were collected in two regions in Ghana, Kenya, Tanzania, and Zambia, in Malawi households were surveyed in four regions and in Mozambique in three regions. Figures 3.2a through 3.2f shows the number and share of FMFs and MMFs respectively by region. The data presented in the tables that follow is presented by region following the same country order (Ghana first and Mozambique last) as found in Figure 3.2.

As can be seen in Figures 3.2a through 3.2f, the number and share of female farm managers varies not only between countries, but this distribution also has a strong regional slant. Four regions have a strikingly high share of FMFs: Kakamega in Kenya, Shire Highlands in Malawi, and the Centre and the South in Mozambique. By contrast, the Upper East in Ghana, Ntchisi in Malawi, Morogoro in Tanzania, and Mazabuka in Zambia have a high share of MMFs.

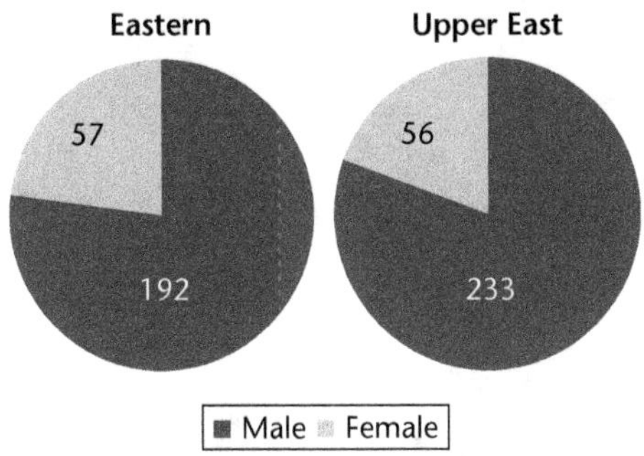

Figure 3.2a Number of male and female farm managers by region, Ghana, Afrint III.

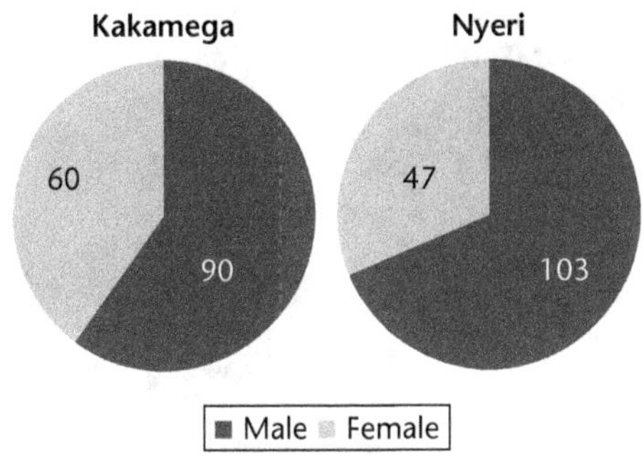

Figure 3.2b Number of male and female farm managers by region, Kenya, Afrint III.

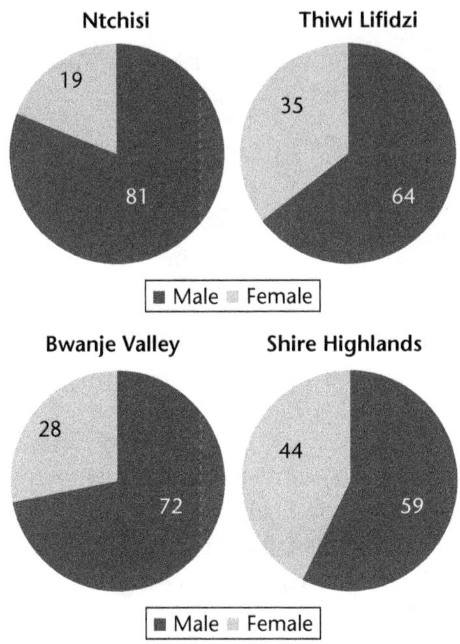

Figure 3.2c Number of male and female farm managers by region, Malawi, Afrint III.

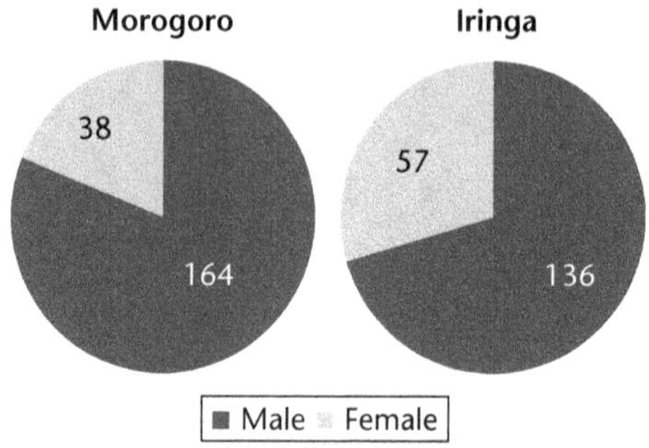

Figure 3.2d Number of male and female farm managers by region, Tanzania, Afrint III.

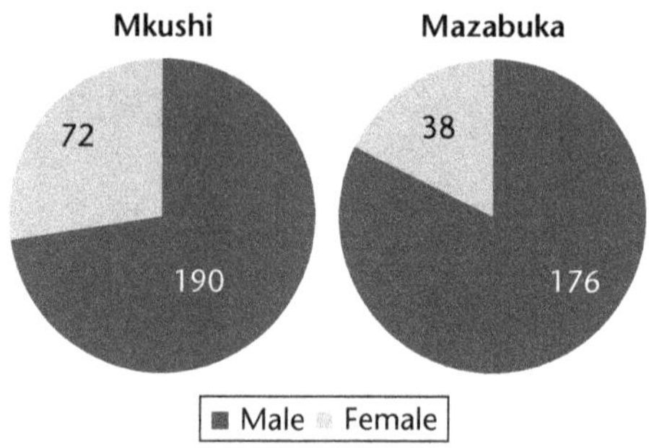

Figure 3.2e Number of male and female farm managers by region, Zambia, Afrint III.

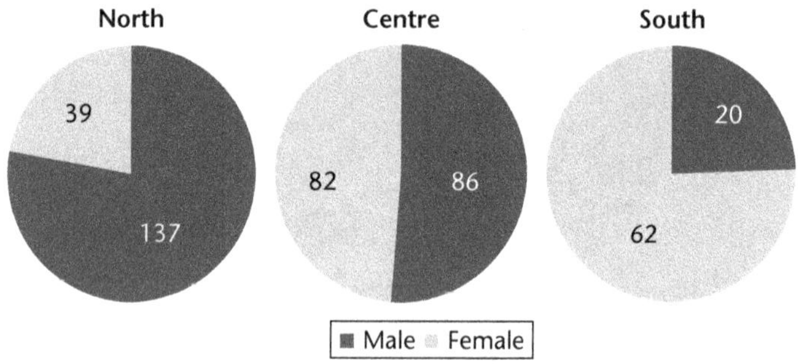

Figure 3.2f Number of male and female farm managers by region, Mozambique, Afrint III.

Table 3.1 shows the average cultivated area by sex of farm manager for the different regions at the three points of data collection. A number of key points can be made on the basis of this table. First, although female farm managers held smaller parcels of land in the majority of regions, there are exceptions, as illustrated by both regions in Kenya and also the South in Mozambique, where for Afrint II, FMFs were larger than MMFs. Second, gender-based differences in

Table 3.1. Mean and median size of cultivated area (ha) by region and sex of farm manager

Country	Region	Afrint I								Afrint II								Afrint III							
		Mean						Median		Mean						Median		Mean						Median	
		Male	N	Female	N	Diff	Sig.	Male	Female	Male	N	Female	N	Diff	Sig.	Male	Female	Male	N	Female	N	Diff	Sig.	Male	Female
Ghana	Eastern	2.73	152	1.60	47	1.12	***	2.30	1.30	1.90	208	1.17	62	0.73	***	1.40	1.00	3.71	188	1.85	57	1.86	***	3.36	1.68
	Upper East	2.15	187	1.76	23			2.00	1.60	2.16	240	1.60	48	0.56	***	1.80	1.40	2.16	227	1.62	56	0.54	***	2.00	1.40
Kenya	Kakamega	1.19	89	1.13	58			1.00	0.93	1.02	91	0.88	47			0.80	0.65	1.05	83	0.89	59			0.92	0.82
	Nyeri	0.67	80	0.64	68			0.60	0.50	0.83	94	0.65	56	0.18	*	0.69	0.55	0.89	102	0.81	47			0.82	0.62
Malawi	Ntchisi RDP	2.04	63	1.53	34	0.51	**	1.82	1.42	1.58	74	1.37	23			1.42	1.01	1.60	78	1.12	18	0.49	**	1.50	1.20
	Thiwi Lifidzi	1.05	61	0.79	30	0.26	**	1.01	0.71	1.28	59	0.98	40	0.30	***	1.21	0.91	1.19	63	1.04	34			1.20	0.90
	Bwanje Valley	1.26	64	1.08	32			1.21	1.01	1.33	58	0.96	40	0.38	***	1.21	0.81	1.25	67	0.86	25	0.39	**	1.20	0.80
	Shire Highlands	0.93	45	0.74	54	0.18	**	0.81	0.66	1.03	29	0.69	68	0.34	***	1.01	0.63	0.94	57	0.63	41	0.31	***	0.80	0.40
Tanzania	Morogoro	1.83	160	1.33	31	0.51	**	1.70	1.20	1.60	170	1.14	31	0.46	***	1.40	0.81	2.14	160	1.52	37	0.62	**	1.77	1.21
	Iringa	2.05	151	1.29	48	0.76	***	1.70	1.00	1.71	142	1.20	42	0.51	***	1.60	1.05	2.04	128	1.31	56	0.73	***	1.72	1.21
Zambia	Mkushi	2.30	134	2.04	74			2.00	1.75	1.95	182	1.34	46	0.62	***	1.74	1.25	2.79	175	1.82	71	0.97	***	2.10	1.40
	Mazabuka	3.85	152	2.24	28	1.61	***	3.50	2.00	3.58	129	1.91	30	1.67	***	3.00	1.44	4.25	169	2.26	38	1.99	***	4.00	1.50
Mozambique	North	2.53	97	1.89	58	0.64	**	2.00	1.50	1.45	130	1.05	24	0.40	**	1.28	0.81	2.28	134	1.62	38	0.66	**	1.87	1.00
	Centre	2.04	90	1.79	56			2.00	1.55	1.00	114	0.79	42	0.21	**	0.88	0.67	3.05	86	2.54	81	0.52	*	3.00	2.00
	South	1.33	32	1.26	43			1.00	1.00	1.00	23	1.26	54	-0.26	*	1.00	1.00	0.76	16	0.80	60			0.70	0.60

Note: a. Extreme cases have been removed by region.

landholding size that emerged in some regions between Afrint I and II have on the whole been retained during the second period. Third, although national data suggest increasing land fragmentation in Kenya and Malawi, only three regions show sizeable, statistically significant decreases in land size between Afrint I and III: Kakamega in Kenya, Ntchisi in Malawi, and the South in Mozambique. While average landholding size fell in all of these regions, only in the case of Ntchisi has it also been accompanied by a gender gap. By the third phase of the project, MMFs were on average nearly half a hectare larger than FMFs in Ntchisi.

While few regions have experienced declines in cultivated area between Afrint I and III, in one region—Bwanje Valley in Malawi—the mean farm size of female farm managers shrunk, but the size of the landholdings of male farm managers were unchanged. In this region, FMFs contracted by around a fifth of a hectare.

In the four regions where land sizes *increased* on average—the Eastern Region in Ghana, Morogoro in Tanzania, Mazabuka in Zambia, and the North in Mozambique—the gender gap in land size widened considerably. This is especially apparent in the Eastern Region in Ghana and Mazabuka in Zambia, where by the time of Afrint III MMFs were on average nearly 2 ha larger than FMFs. A methodological note can also be made at this stage: extreme cases (removed by region) are more likely to have been tied to male-headed households. With the number of extreme cases increasing from thirty to sixty-six cases between Afrint I and III, the gender gap is likely to be underestimated for the latter phase of data collection.

Although the data suggest that gender-based differentials in tenure security and control and use of land are minor, it does also point to clear gender gaps in terms of the number of FMFs and the relative size of holdings under female management. Land size changes have in general also affected FMFs negatively when compared with male landholders. In the four regions where farms have grown for both male and female farm managers the gender gap has widened. A gender gap has, moreover, emerged in one of the three regions where land sizes declined for both male and female farm managers. Furthermore, FMFs have shrunk in one region that shows no overall change in land size. The tendency for declining land sizes to affect FMFs disproportionately is worrying. As suggested by Peters and Kambewa's (2007) work from Malawi, with growing pressure on land, women's rights to land may become more precarious over time.

For land size, moreover, there is a difference between *de facto* and *de jure* female-headed households, with a slightly larger cultivated area among the *de facto* female-headed households for Afrint III: 1.7 ha compared with 1.4 ha for the *de jure* female-headed households (significant at the 5 per cent level).

Labour

Until recently, the point was often made in the literature that in land-abundant Africa, labour—and the control over others' labour—was a more decisive factor in raising agricultural productivity and generating wealth than access to land. More recently such claims have been outpaced by rapidly decreasing land sizes, especially in dynamic and well-placed areas (Jayne et al. 2014). At the same time, an emergent class of middle-size and middle-class farmers is being encouraged by policies in Zambia, Ghana, and Tanzania for instance, prompted also by the dwindling of prospects in Africa's urban areas. As suggested by Table 3.1, land sizes have increased considerably in certain regions and labour in these regions is likely to be a differentiating factor, especially in labour-intensive systems of production. Conversely, shrinking farm size may be compensated for through higher inputs of labour, provided that soil fertility can be maintained in the face of rising population pressure.

Whether gender-based gaps in farm size are compounded by differences in access to labour is a crucial question in the context of widening land gaps. Peterman et al. (2014) review gender differences in non-land agricultural inputs and note that there are comparatively few studies of gender-differentiated access to labour in sub-Saharan Africa. The studies that do exist generally document labour constraints among female-headed households, however.

Horrell and Krishnan (2009) use data from three provinces in rural Zimbabwe and find a significant difference in the availability of working-age adults between male-headed households (4.14 persons) and *de facto* female-headed households (3.12 persons). Given the longstanding history of labour migration in Zimbabwe this is not surprising, however. Although there are few empirical studies of this sort, the received wisdom on studies of female-headed households holds that these households are smaller and contain more dependants than households headed by men (see Chant 2007 for a discussion).

When the full sample is considered, this notion is confirmed: the households of female farm managers have on average 2.8 members of working age compared with 3.4 in households headed by men, thus pointing to a gender gap of 0.6 working-age adults, a difference that is strongly statistically significant. Dividing the data by region suggests that the gender gap in labour resources is concentrated in particular regions, and that such differences vary over time, however. Nonetheless, the number of regions with gender-based differences in the number of working-age adults has increased since Afrint II, with nine out of fifteen regions having such gaps by the final round of data collection. Four regions show a persistent gender gap in labour resources since the second survey—the Upper East Region in Ghana, Shire Highlands in Malawi, Iringa in Tanzania, and Mazabuka in Zambia. In the case of the Upper East and Mazabuka this may be explained by polygamy, and in the case of the Upper East also male southward migration.

Comparing *de jure* and *de facto* female-headed households shows a statistically significant difference of 0.5 working-age adults, with *de facto* female-headed households on average containing 3.2 working-age adults during the last round of data collection, compared with 2.7 for *de jure* female-headed households. *De facto* female-headed households hence appear to be somewhat better placed in terms of labour resources, although less so than male-headed households.

Another persistent claim revolves around the *quality* of labour available in female-headed households, with the presumption often being that these lack access to male labour in particular. In the labour-intensive context of smallholder agriculture this may lower productivity in general on FMFs. Gender-biased taboos and norms surrounding the division of labour, moreover, may delay or even completely exclude the undertaking of certain tasks, thus affecting FMFs negatively (Quisumbing and Pandolfelli 2010, FAO 2011).

The labour of an absent husband can, however, be replaced by the work of other male relatives, for instance adult sons or brothers. With respect to male labour the data suggest that this is not the case, however. The demographic composition of female-headed households is fundamentally different from male-headed households in more or less all regions and countries, showing a persistently lower share of male adults.

As suggested by Figure 3.3, in male-headed households, male family members constitute around half of the adults across the regions. By contrast, the share of male labour in female-headed households was around a third or lower in all regions in Malawi (Ntchisi, Thiwi Lifidzi, Bwanje Valley, and the Shire Highlands) and Tanzania (Morogoro and Iringa), and also in the Eastern Region (Ghana) and the North and South (Mozambique).

De facto female-headed households replicate the patterns of male-headed households. They have a much higher share of male adult labour than the *de jure* female-headed households: indeed at the time of Afrint III, half the adult labour was male in the former compared with 38 per cent for the latter (significant at the 1 per cent level). As pointed out in Chapter 1, *de facto* female-headed households are concentrated to three particular regions characterized by a long history of male labour migration. Qualitative fieldwork from Kakamega in Kenya suggests that the circular mobility of husbands can balance the labour peaks of the agricultural seasons, which may explain the higher availability of labour in *de facto* female-headed households (Andersson Djurfeldt 2012).

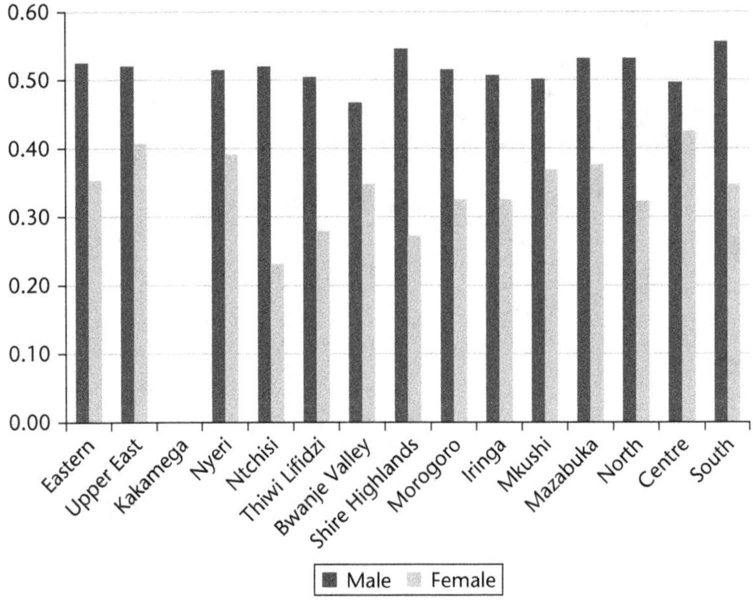

Figure 3.3 Male adult members as share of total household members of working age (16–60), by region and sex of head of household, Afrint III.

Notes: a. The figure shows only regions where there were statistically significant differences between MMFs and FMFs. All levels of statistical significance were below 5 per cent.

In the usufruct land tenure systems found in many of the study sites, the relative lack of male labour may explain smaller farm sizes as well as the more limited ability to produce a marketable surplus (a topic discussed in Chapter 4) among female farm managers. This difference also appears to be reflected in the smaller land sizes of *de jure* female-headed households noted earlier. Where rights to land are based on the ability to use land, access to labour may be important to stake claims to land and also to expand cultivated area. This presumes that labour is the constraining factor on land expansion, however.

During the third data collection round a question was added to address the aspect of labour shortages. While the data in Table 3.2 suggest that female-managed households in general have fewer male members, the data on labour shortages contradict the notion that this leads to labour constraints. Only four regions (both regions in Ghana, Shire Highlands in Malawi, and the Centre in Mozambique) showed gender-based differences in the share of farmers who had experienced labour shortages over the past year. Moreover, in the Eastern Region of Ghana, 48 per cent of the male farm managers experienced labour shortages compared with only 29 per cent of the female landholders. In general, there is a strong geographic component to variations in labour

Table 3.2. Mean and median number of working-age adults by household, by sex of head of household

Country	Region	Afrint I								Afrint II								Afrint III							
		Mean						Median		Mean						Median		Mean						Median	
		Male	N	Female	N	Diff	Sig.	Male	Female	Male	N	Female	N	Diff	Sig.	Male	Female	Male	N	Female	N	Diff	Sig.	Male	Female
Ghana	Eastern	3.8	155	3.6	44			3	3	3.7	213	3.1	63	0.6	*	3	3	3.14	191	3.16	57			3	3
	Upper East	5.0	176	4.9	21			4	4	4.1	243	3.1	48	1.0	***	4	3	4.14	228	3.05	56	1.1	***	4	3
Kenya	Kakamega	4.3	90	3.2	59	1.1	**	3	3	4.8	99	4.6	48			4	4	4.01	90	3.17	60	0.8	**	4	3
	Nyeri	2.5	76	2.4	68			2	2	4.3	97	3.9	56			4	4	3.02	102	2.59	46			2	2
Malawi	Ntchisi	2.6	62	2.1	37	0.5	**	2	2	2.4	77	2.0	23			2	2	2.32	77	1.63	19	0.7	**	2	1
	Thiwi Lifidzi	2.3	64	1.3	34	0.9	***	2	1	2.0	46	2.0	13			2	2	2.19	64	2.06	35			2	2
	Bwanje Valley	2.5	67	2.1	33			2	2	2.4	58	2.1	40			2	2	2.64	70	2.50	22			2	2
	Shire Highlands	2.2	45	1.9	56			2	2	2.4	29	2.0	69	0.4	**	2	2	2.91	56	1.93	41	1.0	***	3	2
Tanzania	Morogoro	3.2	166	2.8	33			2	2	3.0	165	2.7	31			3	2	2.66	159	2.29	38			2	2
	Iringa	2.6	152	2.0	47	0.6	***	2	2	2.7	136	1.9	39	0.8	***	2	2	2.79	134	2.00	57	0.8	***	3	2
Zambia	Mkushi	4.0	135	3.6	75			3	3	4.1	194	4.0	46			4	4	4.09	189	3.26	72	0.8	***	4	3
	Mazabuka	4.7	163	5.3	28			4	6	4.7	130	2.8	31	1.9	***	4	2	4.89	175	3.50	38	1.4	***	5	3
Mozambique	North	3.3	100	2.7	60	0.6	**	3	2	2.4	136	1.5	24	0.9	***	2	1	2.51	132	2.05	39			2	2
	Centre	3.7	94	3.3	65			3	2	3.2	119	2.7	43			3	2	3.99	86	3.45	82			4	3
	South	3.4	33	3.0	45			3	2	3.7	24	2.9	53			3	2	4.00	20	2.66	56	1.3	**	3	2

Notes: a. Extreme cases were removed at the regional level—for Afrint II these cases were heavily concentrated to Thiwi Lifidzi, with forty-one out of fifty-five cases removed in this particular region.
b. For Afrint III, twenty-two extreme cases were removed at the regional level.
c. N equals the number of respondents who answered the questions in that category.

shortages—ranging from 12 per cent of the sample in Iringa, Tanzania stating that they had experienced shortages of labour, to 65 per cent in Ntchisi, Malawi. In turn, this suggests that factors external to the household may be more consequential to labour availability than access to family labour.

Access to Water and Irrigation

As noted by Peterman et al. (2014), few studies exist that quantify gender differentials in water access and use of irrigation, although a large scholarship details the legal, cultural, and practical aspects of women's relatively limited access to water (Zwarteveen 2008). Only two studies related to the gendered use of water in sub-Saharan Africa are found by Peterman et al. (2014): one covers the access to piped water in Limpopo (Hope et al. 2003) and the other the distance to the main irrigation channel among forty-five rice farmers in Benin (Kinkingninhoun-Mêdagbé et al. 2010). Both of these find that women (either women farmers or female-headed households) are disadvantaged in their access to water.

We have aimed to identify a broad range of techniques to capture use and control of water in agricultural production. Information was therefore collected on the access to land that was particularly favourable in terms of natural endowments of water, in addition to data on more technical issues of water control and management. These data are only available for Afrint III.

For the sample as a whole, two types of water sources are the most frequently accessed: 30 per cent of the respondents stated that they cultivate on low-lying swampy land and 29 per cent that they had access to water from a dam, river, canal, or lake and used this water to irrigate crops. A smaller proportion—21 per cent—cultivated on riverbanks or riverbeds. The cultivation either on swampy land or on riverbanks is relatively common in all regions, with a couple of exceptions (Nyeri, Kenya and Mazabuka, Zambia), whereas access to dams, rivers, canals, or lakes is more concentrated geographically. All of these water sources relate closely to bundles of rights that may have different implications for men and women depending on local context.

Gendered inequalities in access to water vary strongly across the regions. Indeed, there are regions (Morogoro (Tanzania), Mazabuka (Zambia), North and South (both in Mozambique)) where no statistically significant differences in access to water sources can be found based on the sex of the farm manager. Nonetheless, in nine of the fifteen regions, male farm managers had superior access to water resources when compared with female farm managers. In two of the regions—Kakamega in Kenya and the Shire Highlands in Malawi—this related both to cultivation on well-watered land as well as to open water sources, suggesting a double exclusion of female farm managers. For the other regions, access to other water sources could somewhat compensate for the gender bias found with respect to a particular source of water.

Further sources of gender bias arise from the use of particular irrigation techniques. In seven of the regions a higher share of male farm managers reported watering their crops through lifting water from open sources. In three of these regions the gender gap was sizeable—a difference of 20 per cent or more in the proportion of farms using water in this way. Lower availability of male labour on FMFs may be affecting female farm managers' ability to use water due to gendered bundles of rights, but can also be connected to the heavy labour involved in irrigating largely through manual lifting and carrying of water.

The largest gap in use of water was found in Shire Highlands (Malawi), where 37 per cent of the male respondents watered their crops through lifting water from a dam, lake, river, or canal, compared with 9 per cent of the female farm managers. Female farm managers in Shire Highlands have poorer access to all types of water sources, pointing to the regionalized ways in which gendered access to and use of particular resources intersect.

The use of irrigation techniques that entail some investment in equipment, for instance water pumps or drip irrigation, is generally less differentiated by gender. There were no gender-based differences in the use of drip irrigation or sprinkler irrigation, while a higher share of male farm managers used water pumps only in two regions. Moreover, in the Centre and North regions in Mozambique and in Ntchisi in Malawi, there were female biases in the use of particular techniques. The explanation for this is related to the extremely low use of any of these technologies: only 2 per cent of the households used drip irrigation, 4 per cent had access to a water pump, and 3 per cent used sprinkler irrigation.

Comparing the two types of female-headed households shows biases towards *de jure* female-headed households with respect to water access, with cultivation on low-lying swampy land or riverbeds and access to water from a dam, pond, or canal being more common among these households. Two irrigation techniques (watering through lifting of water and through the use of a water pump) were also more frequent among *de jure* female-headed households. The explanation for the lower access to water resources and use of irrigation among *de facto* female-headed households seems to lie in the geographical concentration of these households in particular regions where these types of irrigation are less important.

Other Non-Land Assets

As suggested by the regional farm size data, mean cultivated area increased in some regions, while farm sizes fell in others. Under conditions of shrinking farm sizes, the possibility for intensifying agriculture, or at least preventing decreases in soil fertility, is intimately connected to assets that can preserve, if not raise, productivity—either in terms of production volume or the monetary value realized from land.

IMPROVED SEED

Peterman et al. (2014) note that research on inorganic fertilizer, and to some extent improved seed varieties, constitute the majority of studies on gender gaps in input use. Most of these studies find that gender in itself is not a significant explanation for differences in input use, with other factors such as education and wealth being more relevant.

The data contain longitudinal information on the use of improved varieties of maize, sorghum, and rice, as well as the use of inorganic fertilizer. The latter data are of limited value, however, since nearly a fifth (18 per cent) of maize farmers report refraining from the use of fertilizer, as they perceive their soils to be fertile. While data on maize are the most comprehensive—maize being grown by 93 per cent of the Afrint III sample—the relative importance of maize in regional crop patterns varies considerably. To address this variability, and the regional importance of rice and sorghum, I combine the data on improved varieties for the three grain crops (maize, rice, and sorghum), creating a variable showing whether households used or did not use improved seed varieties for any of these crops. This does not detail the amounts used on each farm, but nonetheless provides information on technology use among male and female farm managers respectively.

For the sample as a whole, there were no differences in the use of improved seed varieties related to the sex of the farm manager for Afrint I. Only two regions—Iringa in Tanzania and Mkushi in Zambia—showed any significant differences in seed technology uptake based on gender, with 32 per cent of the male respondents in Iringa using improved seed, compared with 9 per cent of the female farm managers (significant at the 1 per cent level). In Mkushi, by contrast, the gender gap was reversed, with 61 per cent of female farm managers using improved seed varieties compared with only 42 per cent among male farm managers (significant at the 1 per cent level). In general, the interregional variation in the use of improved seed was extremely wide, however, with near universal use of improved varieties in Nyeri (93 per cent) in Kenya and Mazabuka (92 per cent) in Zambia, compared with 1 per cent in the South region of Mozambique.

By the time of Afrint II, gender-based differences in the use of improved seed had emerged for the sample as a whole: 50 per cent of female farm managers used improved seed compared with 54 per cent among male landholders (significant at the 5 per cent level). Breaking down the data by region only shows significant differences based on gender in one region, Ntchisi in Malawi, where 61 per cent of male farm managers used improved seed, compared with

26 per cent among female farm managers (significant at the 1 per cent level). In the meantime regional disparities increased further, with no households using improved seed varieties in the South region in Mozambique, and the share of households using improved varieties in Nyeri rising further to 96 per cent.

The share of respondents in the full sample that used improved seeds fell for both male farm managers (51 per cent) and female farm managers (44 per cent) for Afrint III, while the gender gap widened from 4 per cent to 7 per cent (significant at the 1 per cent level). At the regional level again, however, there was a statistically significant difference only in one region, the Shire Highlands in Malawi, where 80 per cent of male farm managers used improved seed varieties, compared with 57 per cent of female farm managers (significant at the 5 per cent level). Regional differences were again pronounced and on the whole stable over time, with Nyeri (96 per cent) and Mazabuka (95 per cent) having the highest share of households using improved seed, and the regions in Mozambique again having the lowest use of improved seed. In this sense differences in use of improved seed varieties are based on geography rather than gender. There were no differences in the use of improved seeds between the two different types of female-headed households.

AGRICULTURAL TECHNIQUES

In the Afrint II round of data collection, several variables on agricultural techniques were introduced. Unfortunately, this information proved to be difficult to collect and it is not possible to use the data as a baseline. A new attempt made for Afrint III produced more reliable results. While these data cannot provide a longitudinal perspective on the gender dynamics of technology uptake, they can nonetheless shed light on the technology available to male and female farm managers respectively.

Data were collected on a number of agricultural techniques: crop rotation, intercropping use of chemical fertilizer, pesticides, and herbicides, use of animal manure and crop residues, composting, fallowing, no-till agriculture, agroforestry, and techniques used to prevent soil loss due to rain. In general, the use of agricultural technology is quite high, with intercropping and use of inorganic fertilizer being most prevalent, used by 65 per cent and 59 per cent of the households respectively. Zero-till techniques were the least common, but of regional importance in the two regions in Zambia and in the Eastern Region in Ghana (which is also confirmed by qualitative fieldwork carried out in these regions, see Andersson Djurfeldt et al. 2014, Andersson Djurfeldt and Hillbom 2016).

At the overarching level there is a clear gender bias in the use of almost all of the techniques, the exceptions being intercropping and agroforestry. As cautioned earlier, however, production systems as well as gender dynamics differ appreciably across space, with particular techniques being more important to local and regional livelihoods than others. In turn, gender-based access to assets may differ depending on norms that regulate the use of land and commons, for instance practices related to the planting of trees or cultural restrictions surrounding the keeping of livestock.

Running the data by region shows that only one region (the Centre in Mozambique) has no gender gaps in technology use. For the remaining regions, technology gaps that bias MMFs exist for a varying number of technologies, ranging from one to seven techniques. The two regions in Zambia have the largest number of gender gaps in technology use (see Figure 3.4). In the North region in Mozambique there is a female bias in the use of no-till agriculture, but only three female farm managers use this technique so the reliability of the results is questionable in this case.

Even if regional technology use varies noticeably, a few patterns deserve mention: alongside agroforestry, the use of animal manure and pesticide application are the techniques where gender gaps are most prevalent. This confirms broader patterns noted in the literature with respect to gendered patterns of livestock ownership (Kristjanson et al. 2014). The review by Peterman et al. (2014) identifies only two studies from sub-Saharan Africa that consider pesticide use, but out of these a study of 451 household in

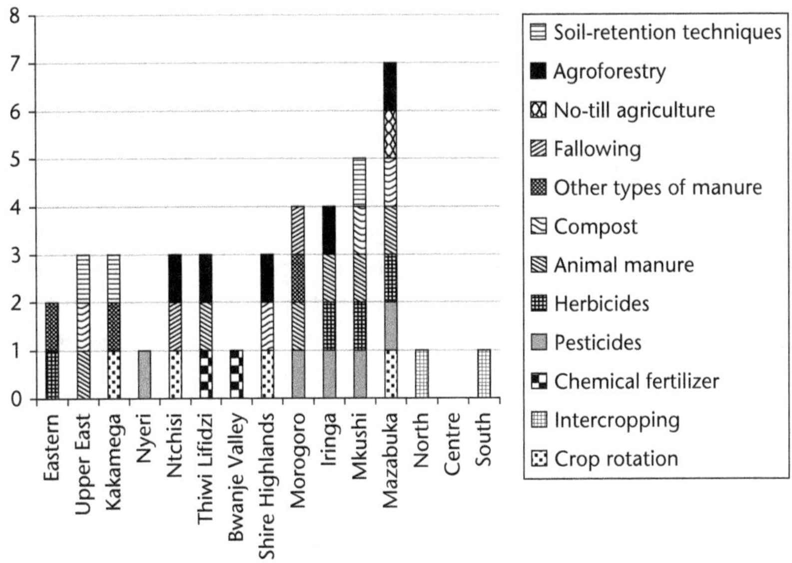

Figure 3.4 Regions with significant differences in gendered use of agricultural techniques.

Notes: a. The share of male farm managers using the technique is higher than the share of female farm managers using the technique, Afrint III.

Uganda (Jagger and Pender 2006) finds no differences in adoption between male- and female-headed households. A smaller study by Kinkingninhoun-Mêdagbé et al. (2010) of forty-five rice farmers in Benin documents gender-based inequalities in adoption of pesticides, however. Drawing on qualitative fieldwork from two villages in Mkushi and one in Mazabuka carried out in November 2012, a number of possible sources of gender gaps in these particular regions spring to mind. Focus group discussions among male farmers suggested that lower literacy levels among women prevents their use of pesticides, since they are not able to read the instructions printed on the pesticide containers, while the equipment used for spraying is too heavy for women to carry. Contract farming arrangements for cotton (a crop which if grown conventionally has high pesticide requirements) meanwhile discriminate against female farmers. Gender-based discrimination may therefore arise both by default as well as by design, depending on the characteristics of cropping systems and use of particular inputs, as well as marketing arrangements (Andersson Djurfeldt 2017).

Agricultural techniques are remarkably similar among *de facto* and *de jure* female-headed households, with a few exceptions: crop rotation and the use of chemical fertilizers are lower among the former, but pesticide use, the use of green manure, and agroforestry were higher among the latter. It appears that these differences are related to the nature of the production systems found in the regions where *de facto* female-headed households are concentrated, however, and it therefore becomes spurious to draw any conclusions on the basis of this variation.

LIVESTOCK

In their review of the literature on gender and livestock, Kristjanson et al. (2014: 210) note a 'dearth of quantitative information on this subject, especially for mixed crop-livestock systems where most livestock and livestock keepers are found'. While the data cannot specify the channels through which women acquire livestock or the extent to which women control the proceeds generated from sale of livestock or animal products, they do enable us to quantify livestock ownership by the sex of farm manager.

A first point relates to regional differences in livestock ownership: four regions stand out in this respect, Mazabuka in Zambia, the Upper East Region in Ghana, and Kakamega and Nyeri in Kenya. In the remaining regions the importance of livestock to livelihoods is more marginal, with households on average holding one tropical livestock unit or less. While gender gaps have widened—especially during the latest period, the size of these gaps are small in regions where livestock are of limited importance, but more sizeable in those regions where livestock ownership is largest. Nonetheless, persistent gender gaps in livestock ownership are found in regions where livestock is relatively less important (Mkushi, Zambia, and Iringa, Tanzania), as well as in the Upper East, Ghana where households on average held 3.5 tropical livestock units by the time of Afrint III. For the first round of data collection there were statistically significant differences in livestock ownership based on the sex of farm manager in five regions, a number that remained unchanged for Afrint II (although the particular regions shifted to some extent) and increased to eight regions by the final survey, as shown in Figure 3.5.

While gender gaps had emerged in numerous regions by the time of the final survey, as illustrated in Figure 3.5, the disparity between MMFs and FMFs is largest in the case of Mazabuka, Zambia. Here female farm managers on average had herds around half the size of male farm managers. In this region there had been continuous restocking following outbreaks of East Coast fever just before the first survey. The average herd size was just above 2 tropical livestock units for Afrint I, increasing to roughly 5 by the time of the second survey round.

Methodological issues related to the number of extreme cases that were removed in the various years suggests that the gender gaps may be underreported in the Afrint III data. Extreme cases were removed at the village level,

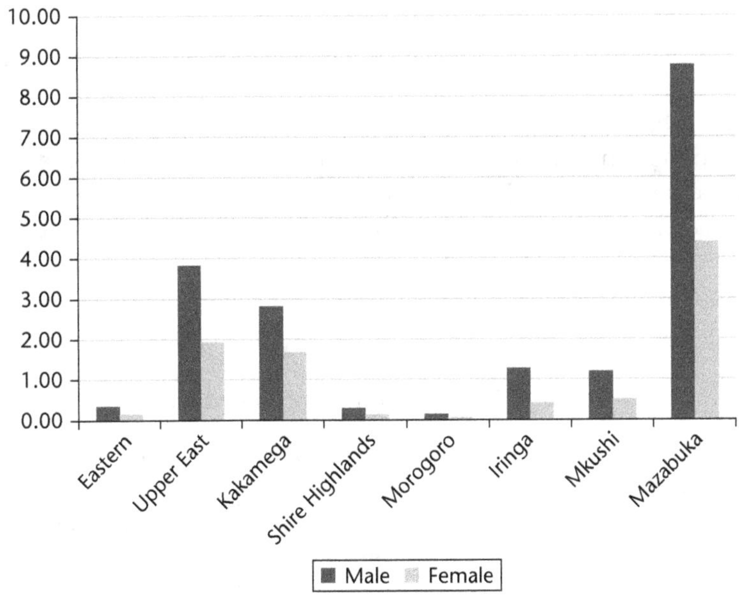

Figure 3.5 Tropical livestock units by sex of farm manager in regions where there were statistically significant differences between MMFs and FMFs, Afrint III.

Notes: a. Eastern and Upper East is in Ghana, Kakamega in Kenya, Shire Highlands in Malawi, Morogoro and Iringa in Tanzania, and Mkushi and Mazabuka in Zambia. The 2013 livestock data for Mozambique are not possible to use, since coding errors in the data prevent comparability with other countries as well as over time.

with around eighty cases removed for the first two rounds, but as many as 119 cases were removed for the Afrint III data. Out of these 101 were MMFs, and only eighteen FMFs.

The assumption is often made that women keep small stock—for instance goats, sheep, and poultry—while men own larger types of livestock, such as cattle. The data from Afrint III modify this impression. In those regions where livestock ownership is high (Mazabuka in Zambia especially, as well as the Upper East, Ghana), male farm managers have higher access to several types of livestock, both cattle as well as small stock. Comparatively, ownership of goats, sheep, and poultry are as biased towards MMFs, as is ownership of cattle. Even if there are examples of female farm managers holding all types of livestock, at the aggregate level there are large gender-based differentials in livestock ownership regardless of animal type. The data therefore suggest that on the whole female farm managers are not able to compensate for lacking cattle ownership through keeping small stock. There are no differences in ownership of livestock among the two types of female-headed households.

Non-Agricultural Assets

While agricultural assets are clearly important in generating income in the context of livelihoods that rely greatly on farming, non-agricultural assets can be used as a basis for diversification away from farming into potentially more lucrative opportunities. Moreover, non-agricultural assets may, in the strongly seasonal context of smallholder-based farming systems, enable offsetting shortfalls in farm-based income and food during the lean season. With respect to gender specifically, access to non-farm assets may also compensate for women's relatively poorer foothold in relation to agricultural assets.

Non-Farm Assets that Can Be Used for Income Generation

Longitudinal data are available for a handful of assets with income-generating potential: data on ownership of mobile or stationary phones, bicycles, and sewing machines have been collected at all three points in time. For Afrint III, data were also collected on the ownership of cars.

The massive expansion of information and communication technologies (ICTs) in rural Africa is reflected in the data: at the start of the project, only 3 per cent of the sampled households reported that they had access to a mobile or stationary phone. By the final round of data collection, 67 per cent stated that they had access to a phone, but regional disparities are wide: only

19 per cent of households in Bwanje Valley (Malawi) possessed a phone, compared with 97 per cent in Nyeri (Kenya). At the country level, Malawi stands out with the lowest share of phone ownership.

The ICT revolution has been largely gender neutral, with some exceptions: in regions with very low rates of phone ownership the gender gaps were pronounced. This characterized three of the four regions in Malawi: in Bwanje Valley, 25 per cent of the male farm managers owned a phone compared with 4 per cent the female farm managers (significant at the 5 per cent level). Similarly, in Thiwi Lifidzi, where only 21 per cent of the respondents owned a phone, there was a large gender gap—27 per cent of the male farm managers reported owning a phone compared with 11 per cent of the female respondents (significant at the 10 per cent level). In Ntchisi, where 30 per cent of all the sampled households owned a phone, the gender gap was 14 per cent—with only 11 per cent of the female farm managers having a phone (significant at the 5 per cent level). Also in the North region in Mozambique, the gender gap was pronounced with 41 per cent of male farm managers owning a phone, compared with 13 per cent of the female farm managers (significant at the 1 per cent level).

For the transportation assets (car, motorbike, and bicycle), only data on bicycle ownership are available for all three rounds, whereas data were collected for car and motorbike ownership only in the third phase of the project. The ownership of cars is extremely limited—only 4 per cent of the sampled households owned a car, with a concentration of car ownership in Nyeri in Kenya and the Eastern Region in Ghana. This variable is therefore not a reliable reflection of income-generating assets for the sampled households. Ownership of motorbikes is restricted to 9 per cent of the sample, preventing any meaningful analysis of gender gaps.

Bicycle ownership is more inclusive—already during Afrint I, half of the sample reported owning bicycles, a figure that increased to 60 per cent by the second survey and remained at that level until the final round of data collection. For the sample as a whole, there are large and growing gender-based differences in bicycle ownership: for Afrint III, 69 per cent of male farm managers had a bicycle, compared with 39 per cent of female farm managers (significant at the 0.1 per cent level). Gender-based differences in bicycle ownership are large and widespread, and found in all regions except one (the Upper East in Ghana). This may be an indicator both of a generally lower asset base among female farm managers, but could also be suggestive of the more restricted mobility of women. A final asset that can enable raising non-farm income is the sewing machine. Again, however, the ownership of sewing machines is limited, with only 9 per cent of the households having a sewing machine by the time of Afrint III.

Housing Standard, Other Consumer Durables, and Savings

While the consumer durables discussed in the preceding section can be used to improve livelihoods, others may be indicative of the accumulation of wealth (and in that sense they can be seen as livelihood outcomes).

One obvious indicator of wealth is housing standard, for which we have data since the start of the project. The share of households living in 'advanced housing', defined as block and brick houses with corrugated iron roofs, has been increasing steadily, from 18 per cent in Afrint I to 36 per cent in Afrint III. Whereas this type of housing was heavily concentrated to just three regions (Nyeri in Kenya, Morogoro in Tanzania, and Eastern Region in Ghana) in the first round, by the time of the last survey more than 20 per cent of the households in ten regions lived in advanced housing. Gender gaps in housing standard were generally absent—only in four regions, Mazabuka and Mkushi in Zambia, Iringa in Tanzania, and the Shire Highlands in Malawi did female farm managers have a poorer housing standard than their male counterparts. Moreover, in two of these regions, statistical significance was only possible to establish at the 10 per cent level.

In addition to housing standard, we have data on four other consumer durables, as well as the ability to save (whether the household was able to save or not). Longitudinal data are available for ownership of TVs and modern stoves, while this information was complemented by data on sofa sets during the final round of data collection. Data on savings are available for all three surveys. In general, ownership of both TV sets and modern stoves was sparse for the first two rounds of data collection (between 10 per cent and 15 per cent of the households had these assets), with the possibilities for analysing gender gaps being limited by small sample sizes. By the time of Afrint III a quarter of the sample owned a TV, but these households were strongly concentrated to the regions of Nyeri (71 per cent) in Kenya, Mazabuka (50 per cent) in Zambia, and the Eastern Region (41 per cent) in Ghana, and to a lesser extent to the other two regions in Zambia (Mkushi, 35 per cent) and Kenya (Kakamega, 30 per cent). The largest gender gap existed in Mkushi, where 43 per cent of the male respondents had a TV, compared with only 14 per cent of the female farm managers. Kakamega and the Eastern Region had no gender gaps in TV ownership. The data on sofa sets replicate the data on TV ownership—Eastern Region, the two regions in Kenya, and the two regions in Zambia have the highest share of households with sofa sets, but for this variable there was a gender gap only in the two Zambian regions.

The longitudinal data on saving ability show that this generally increased during the period, with 42 per cent of the households stating that they could save for future needs at the start of the project, compared with 55 per cent by the time of the final data collection round. During the same period the gender gap widened from 10 per cent to 14 per cent (all differences significant at the 1 per cent level). Breaking down the data by region suggests a different picture, however. In general, during the first phase, the ability to save was much less regionally concentrated than data for consumer durables at the corresponding point in time, from a low in Thiwi Lifidzi in Malawi, where 22 per cent of the households reported that they could save, to 58 per cent in the Eastern Region, Ghana. Moreover, there were significant gender differentials in the ability to save in only three regions. By the time of the second survey, gender gaps in saving ability (biased against female farm managers) were found in six regions, pointing to increasing gender-based differentiation amid general improvements in the ability to save. Interestingly, however, in the case of Kakamega, Kenya, savings ability was higher among female farm managers than among male farm managers (a gap of 19 per cent).

During Afrint III, somewhat surprisingly, the lowest ability to save was found in Kakamega, where 29 per cent of the households stated that they saved, whereas the two regions in Zambia (where 83 per cent and 72 per cent of the respondents saved), Nyeri (71 per cent) in Kenya, and the Eastern Region (71 per cent) in Ghana held the largest share of savers. Five regions had significant gender-based gaps in the ability to save. Overall, the data on savings suggest improvements over time, with regional disparities closing and gender gaps being maintained only in a handful of regions.

Advanced housing as well as consumer durables were more common among the *de facto* female-headed households than the *de jure* ones—but again, this may be related to regional effects, where the concentration of consumer durables in the two Kenyan regions overlaps with a high share of *de facto* female-headed households. There were no differences in the ability to save.

Conclusions

A few initial observations can be made regarding the trends in the data. Sources of gender-based asset gaps vary regionally—most likely depending on local production systems. Moreover, gender gaps shift over time—there are few persistent trends in the data stretching as far back as the first round of data collection, although in several regions gaps have emerged between Afrint I and II that have been sustained or widened during the latest period of data collection.

With respect to particular asset types, agricultural assets, especially land and water, were generally biased against female farm managers. Female farm managers were fewer than male farm managers in all regions and on average held smaller farms than male farm managers in eleven out of the fifteen regions. Moreover, changes in land size—whether positive or negative—had a disproportionately negative effect on FMFs when compared with MMFs.

Somewhat surprisingly, however, control over land was not biased against female farm managers—indeed a slightly higher share of these respondents stated that they had full control over land, compared with male farm managers. Moreover, female farm managers had not experienced loss of land to a greater extent than their male counterparts. Gender biases with respect to land appear therefore to lie primarily in the sizes of cultivated areas, which in usufruct systems is related to the ability to control labour.

The demographic data show few statistically significant differences when it comes to the size of the households managed by women and men respectively, but the share of male labour is much lower on FMFs across all regions, except one. In turn, this appears to be connected to smaller land sizes as well as the lower use of particular irrigation techniques that depend on the manual lifting of water. While the importance of livestock varies strongly between the regions, in those regions where livestock are most important, the gender gaps are also pronounced. Moreover, in these regions male farm managers dominate all types of livestock ownership. Overall, therefore, the results suggest that the ownership or access to agricultural assets are in general concentrated in MMFs, with the importance of particular types of capital varying depending on the characteristics of regional production system.

In general, the data on housing standard, consumer durables, and savings suggest that assets that capture livelihood outcomes are less prone to gender-based differentials than agricultural assets that enable raising incomes and securing livelihoods. The explanation for this discrepancy may lie in the differences in household size—since female farm managers in general command less male labour the land that they cultivate and the livestock herds that they keep appear to be adjusted to their labour resources. Incomes generated (whether from agriculture or not) by these households are invested in improved housing, consumer durables, and savings, rather than used to accumulate agrarian assets. In terms of policy recommendations this suggests that labour-saving technologies and crops may be more important interventions for FMFs than labour-intensive packages focused primarily on raising yields. Moreover, the tendency among female farm managers to invest in non-agricultural assets also points to the need for encouraging income opportunities related to such assets, for instance the possibility of engaging in local processing of agricultural products to add value.

4

Gender and Rural Livelihoods: Agricultural Commercialization and Farm/Non-Farm Diversification

Agnes Andersson Djurfeldt

Introduction

As noted in Chapter 1, the smallholder-based rural development model strongly emphasizes commercialization. In this sense, linking small farmers to agricultural output markets—whether domestic or global—constitutes a pivotal aspect of strategies aimed to encourage pro-poor agricultural growth.

The possibilities of linking smallholders to markets rest on supply-side as well as demand-side factors, related to the production of a marketable surplus as well as the functioning of markets in themselves. Much research has been concerned with the gendered patterns of agricultural production, tied to aspects such as cropping patterns (Doss 2002, Carr 2008), technology uptake (Doss and Morris 2001, Doss 2001, Peterman et al. 2010, Quisumbing and Pandolfelli 2010, Bergman Lodin 2012), gender gaps in productivity, and access to extension services (Kilic et al. 2015). With the exception of the global value chain literature (Barrientos et al. 2003, Maertens and Swinnen 2009), much less interest has been devoted to the gender dynamics of commercialization, both in the mainstream developmentalist literature and in feminist post-structuralist work.

Gender-based studies of rural livelihoods, meanwhile, tend to depart from land and labour *markets*, whether dealt with through the economic perspectives of allocative inefficiencies (Udry et al. 1995, Udry 1996) or as seen through the structuralist lens of political economy (Razavi 2009), rather than the empirical question of how women and men make a living in rural areas. Given the well-known male bias in land tenure systems in most of

Africa, the relative discrimination of women under these tenure systems is by now well documented (Peters 1997, FAO 2010, Peters 2010, FAO 2011, Peters 2013). Less is known, however, regarding how, or even if, such biases translate into exclusion from agricultural output markets too, and how this in turn may be connected to women's access to non-farm sources of livelihood.

This chapter uses a cross-country comparative perspective in analysing gendered patterns of agricultural commercialization and rural livelihoods. A first research question addresses whether female farm managers are in fact excluded from agricultural commercialization (and by implication incomes) when compared to their male counterparts. Whether the sources of this exclusion can be found in the functioning of markets themselves or factors inherent to the household constitutes an important sub-question. Second, the chapter analyses if and how access to non-farm incomes (NFIs) varies by gender, and by extension whether incomes from the non-farm sector can compensate for poorer access to agricultural incomes among female farm managers. Third, how the prospects for commercialization and livelihood diversification among the two different types of female-headed households (*de facto* and *de jure*) vary will be considered. Finally, the income-generation patterns of those women who live in male-headed households will be addressed. The analysis in what follows will be guided by these questions, and positioned in relation to existing theoretical and empirical research frontiers and gaps.

Theoretical Points of Departure

As suggested initially, two main tenets underpin pro-poor growth approaches to agricultural development in sub-Saharan Africa: inclusivity and commercialization. In the grey area literature, prospects for enhancing inclusivity among female farmers are centred on closing the gender gap in agriculture (World Bank et al. 2009, FAO 2011) through redressing productivity differentials arising from differences in access to productive resources and non-land inputs.

Policy perspectives in turn draw on a vast literature on gendered access to agrarian resources, especially with respect to factor markets, both among agricultural economists in what is sometimes referred to as the mainstream feminist literature (Doss and Morris 2001) as well as political economy approaches (Jackson 2003, Razavi 2003, Razavi 2009, Peters 2013). While the latter bring to the fore issues of power, social relations, and historical change that are generally absent in the former, both bodies of scholarship demonstrate a dividing line in the command over agrarian resources based at least in part on gender.

More surprising, given the aspirations of pro-poor growth strategies to 'link smallholders to markets' (Bernstein and Oya 2014), is the limited literature available on gendered aspects of commercialization, especially for food crops—crops that (often mistakenly) are assumed to be 'women's crops' (Doss 2002, Carr 2008). While women's relatively limited commercialization at one level is strongly connected to the productivity constraints identified in the literature, situating commercialization in relation to broader livelihood aspects can shed important light on commercialization dynamics, especially with respect to food crops.

Agricultural Commercialization and Gender

In a recent review of studies of market engagement among African smallholders, Wiggins and Keats (2013) identify a number of explanations for the limited commercialization among smallholders in general. Ill-fitting technologies and high transportation costs suppress potential marketable surplus or restrict the profitability of trading. High-risk environments characterized by insecure rights of tenure and erratic government policies supress commercialization; credit constraints and inflated transaction costs provide further explanations. Monopsonistic output markets in some countries lower producer prices, hampering the incentives for selling produce.

Out of these aspects, the ones that have received the most interest in the literature on gender and commercialization are farm-based constraints related to technology uptake, extension services, and insecurity of tenure, tied explicitly or implicitly to lower productivity and the poorer possibilities for the generation of a marketable surplus.

Beyond the supply-side factors connected to poorer female productivity, two main bodies of literature have emerged in relation to women's commercialization in sub-Saharan Africa. On the one hand numerous studies, both contemporary and historical, concern the male dominance over cash crop production and sales as well as the growing male control over women's crops as they become commercialized (Von Braun and Webb 1989, Moore and Vaughan 1994, Lilja et al. 1996, Sørensen 1996, Kasente et al. 2002, Negin et al. 2009, Bergman Lodin 2012).

A second body of literature departs from (global) value chain analysis and concerns gendered aspects of various value chains as well as the gendered outcomes of different types of marketing arrangements, for instance cooperatives or farmers' groups (see Rubin and Manfre 2014 for a recent review) in such value chains. The bias against women in terms of their relatively limited access to productive assets, technology, credit, information, and social networks explain their inability to engage in value chains of higher-value or contract farming arrangements (Maertens and Swinnen

2009, Schneider and Gugerty 2010). Demands to meet increasingly taxing food safety and quality standards, and the inability to fulfil the volume requirements of large-scale buyers, raise the entry costs for high-value markets in particular (Reardon et al. 2009).

Studies concerned with the gender dynamics of market coordination arrangements that enable access to value chains are limited. A study from Tanzania suggests that farmers' marketing groups dominated by women are less successful in accessing markets than male groups, although the reasons for this, as cautioned by the authors, may well rest with the lack of access to natural resources (Barham and Chitemi 2009). Fischer and Qaim (2012) show how collective action among Kenyan small-scale banana farmers is tied to increased male dominance over revenues generated from sales, but also that female membership in farmers' groups can counteract these tendencies. Intra-household aspects can prevent women in male-headed households from joining marketing cooperatives however, as members need to overcome 'lack of trust and suspicion concerning their whereabouts' and the husbands' fear of 'losing control over them, restricting their engagement in groups' (Gotschi et al. 2009: 272), as noted in the context of Mozambique. Finally, Handschuch and Wollni (2013) identify a general empirical gap in relation to the marketing of food crops, while showing the potential of collective action among female farmers of finger millet in Kenya to increase market participation as well as prices.

Complementarities and Commercialization

While studies of value chains, commercialization, and market participation focus on particular crops, complementary perspectives are concerned with shifts in intra-household relations and production dynamics connected to processes of commercialization. In this vein, studies of intra-household division of labour show how women withdraw their labour from production of crops whose commercialization predominantly benefits their husbands, to engage instead in food crop production or non-farm activities (Whitehead and Kabeer 2001). In this context it can be noted that a common argument for enhancing women's participation in output markets relates to the perceived potential for augmenting independent female incomes that in turn benefit women through increasing their bargaining power within households (Doss 2013, Doss 2014, Rubin and Manfre 2014).

Political economy perspectives contextualize households and intra-household gender dynamics, stressing the joint as well as conflicting activities, interests, and responsibilities of wives and husbands over the life cycle (Jackson 1999, Jackson 2007, O'Laughlin 2007, Okali 2012). The necessity of placing the household in a broader economic as well as institutional context is highlighted by these perspectives. The contextual nature of both gender relations and production systems also raises the need for geographical contextualization.

Rural Livelihoods, Gender, and Non-Farm/Farm Interaction

In this respect, studies of diversification (and livelihood perspectives more broadly) offer a valuable contribution to studies of gender and agricultural commercialization. As noted earlier, the sources of female exclusion from agricultural livelihood opportunities are manifold and varied, involving limitations in the access to productive assets and less tangible resources, as well as restrictions tied to lacking individual mobility and time poverty. Understanding if and how this exclusion interacts with a marginalization from non-farm activities, or may indeed be compensated for through engagement in non-agricultural pursuits, is crucial to analysing gender relations in rural areas and may also shed light on gendered patterns of commercialization.

Nonetheless, Whitehead and Kabeer (2001: 13), summarizing the literature from the 1990s, note that 'there is little systematic research on women's non-farm income activities in sub-Saharan Africa, but what data there is testifies to their active involvement'. As suggested by a recent review by Alobo Loison (2015), numerous studies document the segmentation of non-farm activities into low-entry low-return and high-entry high-return activities respectively (Barrett and Bezuneh 2001, Marenya et al. 2003, Reardon et al. 2007). Only a limited number of studies analyse the gender dynamics of this segmentation (Canagarajah et al. 2001, Lay and Mahmoud 2008), and even fewer consider the gendered ways in which farm- and non-farm activities are linked in rural livelihoods (Yaro 2006, Andersson Djurfeldt et al. 2013).

Data Sources, Reliability, and Analysis

The analysis in this chapter relies heavily on the quantitative data as described in Chapter 1, with qualitative data from three villages in Malawi and Zambia respectively (collected in 2012), two Kenyan villages (collected in 2007), and four villages in Ghana (collected in 2011) being used mainly to illustrate and explain the tendencies identified in the quantitative data. The data collection methods are described in detail in Chapter 1.

Variables are used that concern three aspects of rural livelihoods in particular: commercialization, cash incomes, and cash income composition. For maize—the only grain staple that is grown across the six countries—commercialization is measured through market participation, share of produce sold, and volume sold for the three data collection rounds, by sex of farm manager. For other food crops, data are only available for market participation, and for cash crops only data on whether the crop was grown are available (with the presumption that cash crops are always sold). Production and commercialization data for particular crops are only available at the farm level.

For cash income composition, however, two sources of data are available for the Afrint III data: first, the farm manager was asked to estimate the size of cash income for the household as a whole; and second, individual cash income data, by source, were collected for all adult household members by sex within each household. Although the lack of individual data for Afrint II prevents outright comparison, differences in size and composition of cash income by sex of farm manager can be traced over time, while gender patterns of intra-household sources of livelihoods can be described for the final round of data collection at least.

The drawbacks of relying on cash income data to analyse livelihoods in contexts where subsistence production is widespread were acknowledged in Chapter 1. Nonetheless, the varied nature of production systems across regions and countries, and the lack of detailed production data for roots and tubers, prevent the calculation of total household income.

Statistically the analysis is descriptive: data on commercialization and incomes are compared by sex of farm manager, with differences of means between the two groups analysed through ANOVA (analysis of variance) tests. Given the risk of type 1 errors connected to the high number of statistical tests carried out, differences that are statistically significant above the 1 per cent level need to be interpreted very cautiously.

Agricultural Commercialization and Gender

Patterns of commercialization for maize, non-grain food crops, and cash crops are analysed in what follows. While the dataset also contains information on rice and sorghum, maize is the only grain crop that is grown across all of the countries, although its national and regional importance varies greatly.

Maize

Even at the national level and for one particular crop (in this case maize) it is difficult to draw any clear-cut conclusions with respect to commercialization or gendered patterns of the same.

COMMERCIALIZATION TENDENCIES

Whereas food crops such as maize are frequently described as 'women's crops' (Sachs 1996, Arndt and Tarp 2000), commercialization in terms of market participation in maize is widespread among both male and female farm managers, especially in Tanzania and Zambia. While market participation was biased towards male farm managers in Ghana and Tanzania at the time of Afrint I, this bias disappeared by the second survey, with market participation for maize generally falling for both household types in Ghana since the start of the project. In the case of Malawi, market participation has increased for male farm managers, but decreased for their female counterparts since Afrint II.

By contrast, commercialization by all measures has increased for both male and female respondents, especially in Zambia between the third and second rounds of data collection, while marketed volumes were higher in Ghana for both household types compared to Afrint II (see Table 4.1).

Maize commercialization in Zambia was disproportionately favouring male farm managers already by the second survey, with 61 per cent of maize-growing households in this category participating in the market compared to 46 per cent of the female farm managers. This gap in market participation had expanded further by the time of Afrint III, although both groups of households became increasingly commercialized. Whereas the gap in share marketed was largely similar for the second and third phase, again the share marketed had increased for both household types by the time of Afrint III, with male-headed households marketing 56 per cent of their production, compared with 46 per cent for the female-headed households. The disparity in *amounts* sold was also much larger by this time, suggesting that the sources of increased differences in commercialization can be traced to differences in marketable surplus rather than market segmentation or differences in marketing behaviour as such.

The demographic data discussed in Chapter 3 point to lower access to male labour on FMFs. Qualitative fieldwork carried out in Zambia and Malawi also confirms the notion of lower labour availability in female-headed households. In the case of two of the Zambian study sites (Chilekwa and Nikantaka) visited in 2012, the introduction of conservation farming in 2010/11, as well as the reintroduction of fertilizer subsidies, were perceived to be sources of increased production and yields as expressed in individual as well as group interviews. Conservation farming was being promoted by an NGO, the Conservation Farming Unit (CFU), in collaboration with the Ministry of Agriculture and was based on the construction of basins, which was perceived to be very labour intensive. While representatives from the CFU in the two sites suggested that female farmers were more actively embracing conservation farming the labour intensity of conservation farming adds to the workload of women. Lack of labour resources was identified by female focus groups as the major constraint to production on FMFs. As noted by Haggblade et al. (2010), the labour intensity of conservation farming falls over time but is high in the short term as basins are constructed.

Table 4.1. Commercialization in maize, by country and sex of farm manager

		Afrint I						Afrint II						Afrint III					
		Male	N	Female	N	Diff.	Sig.	Male	N	Female	N	Diff.	Sig.	Male	N	Female	N	Diff.	Sig.
Ghana	Market participation	0.97	156	0.87	46	0.10	**	0.67	296	0.74	73			0.58	337	0.54	81		
	Share sold of production	0.61	150	0.59	40			0.65	199	0.68	54			0.57	197	0.54	44		
	Sale of maize (kg)	481	142	337	39	143	**	516	190	380	54	137	**	655	187	420	44	234	***
Kenya	Market participation	0.21	171	0.16	128			0.31	196	0.33	104			0.34	193	0.34	107		
	Share sold of production	0.39	36	0.34	20			0.51	60	0.46	34			0.44	65	0.38	36		
	Sale of maize (kg)	513	34	513	20			588	57	315	33	273	**	687	63	440	35		
Malawi	Market participation	0.07	239	0.03	160			0.39	224	0.33	171			0.41	275	0.27	125	0.14	***
	Share sold of production	0.17	16	0.09	5			0.25	85	0.20	56			0.22	113	0.24	34		
	Sale of maize (kg)	167	14	81	5			300	79	206	55	94	*	267	103	220	33		
Tanzania	Market participation	0.55	286	0.34	70	0.21	***	0.48	294	0.47	64			0.57	275	0.59	82		
	Share sold of production	0.41	155	0.35	24			0.46	128	0.43	29			0.43	156	0.50	47	−0.07	*
	Sale of maize (kg)	570	150	384	23			711	121	489	29			892	152	695	46		
Zambia	Market participation	0.24	270	0.25	92			0.61	354	0.46	89	0.15	**	0.79	366	0.59	108	0.20	***
	Share sold of production	0.35	65	0.33	23			0.49	215	0.41	41	0.08	**	0.56	290	0.46	64	0.10	***
	Sale of maize (kg)	524	59	324	23	200	*	1,880	202	1,079	41	801	**	2,926	275	1,175	63	1,750	***
Mozambique	Market participation	0.36	197	0.27	145	0.09	*	0.33	231	0.17	113	0.16	***	0.26	229	0.27	174		
	Share sold of production	0.42	70	0.40	39			0.36	77	0.20	19	0.15	***	0.30	57	0.22	46	0.08	*
	Sale of maize (kg)	218	67	198	39			269	75	136	19	133	*	226	51	197	45		

Notes: a. Market participation: share of producing households stating that they sold maize.
b. Share sold of production: share of total production sold.
c. Sale of maize (kg): total amount of maize sold.
d. Extreme cases for amounts of maize sold have been removed at the village level.

The ability to make full use of conservation farming techniques may therefore be limited by labour shortages on FMFs, especially in the short term (Andersson Djurfeldt and Hillbom 2016). The shortage of labour on FMFs was also identified as the distinguishing feature of female farm managers in the focus group discussions across the three Malawian villages where fieldwork was carried out in 2012. The individual interviews with women who were heading their own households also attributed the loss of male labour as a result of divorce, separation, or widowhood to a gradual loss of productivity, as soil fertility was not possible to maintain over time.

MARKETS AND PRICES

Maize markets are generally not physically segmented by gender—maize was sold exclusively within the village by both male and female farm managers in all countries, except for Zambia. In Zambia, by contrast, 47 per cent of the male farm managers sold maize within the village only, compared to 70 per cent of the female farm managers (a difference that is statistically significant at the 1 per cent level). Most farmers (79 per cent) in the Zambian sample rely on the state marketing board as their main marketing channel, however, with no statistically significant differences between MMFs and FMFs, suggesting that market accessibility is not differentiated by sex, even if the location of sales differs.

Further support for the interpretation that constraints to women's commercialization in maize can be found mainly in relation to factors endogenous to the households, rather than in the discrimination of women in output markets, is provided by price data on average maize prices received. Table 4.2 shows that in general prices were the same for both male and female farm managers, with two exceptions—in Ghana for the second round of data collection, the average price received by the latter were actually higher than the former, while the reverse was the case in Malawi. By the time of Afrint III these differences had disappeared, however.

Table 4.2. Mean price of maize per 100 kg received, Afrint II and III (2010, PPP-adjusted USD), by sex of farm manager

	Afrint II						Afrint III			
	Male	N	Female	N	Diff.	Sig.	Male	N	Female	N
Ghana	65	195	71	52	−6	*	66	195	66	44
Kenya	57	58	52	33			58	62	55	34
Malawi	39	75	36	42	3	*	22	107	22	31
Tanzania	47	132	48	26			28	153	28	47
Zambia	40	173	38	27			36	277	35	61
Mozambique	32	77	28	19			40	49	39	37

Note: a. Extreme cases have been removed by region.

Even if price discrimination does not appear to be forthcoming, qualitative data from the three villages in Malawi illustrate the limitations to women's mobility when compared with men. Here individual interview data with both spouses as well as group interviews with women show how female mobility is restricted by domestic chores, especially related to caring for young children and family members suffering from ill-health. While the data underscore the need for contextualization they also point to gendered differences in the ability to participate in markets, related primarily to labour constraints and the lack of marketable surplus.

Non-Grain Food Crops

The gender dynamics of both production and commercialization change over time and are also to some extent country specific, as suggested by the data on non-grain food crops. A few trends are suggested by the data in Table 4.3: first, at a general level, despite the large number of crops covered, there are few strongly significant differences in market participation based on the sex of the farm manager; and second, patterns shift over time. For Ghana, a movement into markets for plantains and out of groundnuts can be noted since the first survey, with no statistically significant differences in market participation based on the sex of farm manager. In Kenya, Irish potatoes have become increasingly commercialized during the same period, again without any gendered biases in production or market participation. For sweet potatoes and millet, moreover, market participation is skewed towards FMFs with male market participation declining and female market participation increasing for these crops since Afrint I. In turn this suggests a feminization of these particular markets over time.

In Malawi, male farm managers have increased their participation in the sweet potato market relative to female farm managers. Market participation in cassava meanwhile has increased disproportionately among male farm managers, while both female and male farm managers have withdrawn from vegetable markets. Although the production of vegetables has increased among both household types since Afrint II, market participation has declined (significant at the 1 per cent level) and a male-biased gender gap in market participation has emerged in the process. In Tanzania, a gender-based difference in the market for beans has re-emerged since the first phase of the project, with male farm managers having significantly higher market participation than their female counterparts by the time of the concluding survey. Meanwhile the female biases in three crops—peas, yams, and other food crops—have disappeared. Here there seem to be some signs of the displacement of female farm managers in these markets, as market participation of male farm managers has increased at the same time. In Zambia, despite the large

Table 4.3. Market participation for other food crops by sex of farm manager (share of growers participating in particular markets)

	Crop	Afrint I						Afrint II						Afrint III					
		Male	N	Female	N	Diff.	Sig.	Male	N	Female	N	Diff.	Sig.	Male	N	Female	N	Diff.	Sig.
Ghana	Cassava	0.95	157	0.85	46	0.10	**	0.90	206	0.92	61			0.88	199	0.83	54		
	Plantains	0.47	47	0.67	12			0.79	154	0.73	40			0.87	158	0.88	43		
	Beans	0.39	188	0.42	24			0.28	226	0.40	35			0.31	218	0.40	48		
	Peas	0.20	10	0.50	2			0.07	15	0.00	3			0.21	24	0.29	7		
	Irish potatoes	0.50	2					0.00	4					0.00	2	0.00	1		
	Sweet potatoes	0.40	105	0.56	9			0.25	72	0.00	3			0.29	73	0.17	6		
	Millet	0.26	172	0.20	20			0.06	142	0.05	22			0.10	228	0.06	54		
	Groundnuts	0.62	217	0.54	28			0.45	209	0.42	31			0.31	232	0.33	48		
	Yams							0.60	129	0.45	29			0.58	120	0.48	31		
	Cocoyams							0.80	165	0.87	38			0.80	153	0.79	38		
	Arrowroot																		
	Vegetables	0.62	236	0.64	45			0.76	268	0.78	60			0.62	288	0.50	70	0.12	*
	Fruits							0.59	37	0.67	3			0.81	47	0.67	9		
	Other	0.56	102	0.78	18	−0.22	*	0.53	17	0.60	5					0.00	1		
Kenya	Cassava	0.25	4	0.50	2			0.05	65	0.09	35			0.20	83	0.31	51		
	Plantains	0.31	133	0.24	99			0.29	173	0.27	89			0.31	159	0.40	89		
	Beans	0.28	165	0.29	123			0.15	180	0.14	92			0.37	191	0.39	104		
	Peas	0.48	61	0.34	50			0.18	51	0.18	28			0.53	68	0.48	25		
	Irish potatoes	0.43	88	0.54	81			0.36	96	0.39	54			0.58	113	0.58	50		
	Sweet potatoes	0.35	124	0.26	94			0.26	140	0.29	63			0.26	149	0.39	79	−0.13	**
	Millet	0.29	17	0.23	13			0.15	26	0.29	14			0.22	32	0.48	21	−0.26	*
	Groundnuts	0.25	40	0.17	23			0.12	42	0.15	26			0.22	32	0.16	19		
	Yams							0.14	37	0.19	16			0.19	26	0.14	14		
	Cocoyams							0.13	8	0.00	1								
	Arrowroot													0.04	69	0.20	30		
	Vegetables	0.69	162	0.60	111			0.76	157	0.64	69	0.12	*	0.56	176	0.58	93	−0.02	**
	Fruits							0.34	74	0.34	32			0.37	153	0.38	81		
	Other	0.25	50	0.16	45			0.28	25	0.20	15			0.20	5	0.33	3		

(continued)

Table 4.3. Continued

	Crop	Afrint I						Afrint II						Afrint III					
		Male	N	Female	N	Diff.	Sig.	Male	N	Female	N	Diff.	Sig.	Male	N	Female	N	Diff.	Sig.
Malawi	Cassava	0.32	56	0.25	44			0.35	69	0.28	60			0.45	75	0.18	39	0.27	***
	Plantains	0.33	88	0.41	59			0.65	46	0.44	48	0.21	**	0.50	105	0.49	41		
	Beans	0.22	139	0.28	94			0.43	134	0.55	110	−0.12	*	0.64	151	0.55	74		
	Peas	0.21	63	0.26	43			0.38	37	0.51	55			0.40	100	0.27	48		
	Irish potatoes	0.57	51	0.32	19	0.25	*	0.72	50	0.52	23			0.80	56	0.79	19		
	Sweet potatoes	0.19	151	0.12	88			0.41	91	0.32	73			0.52	158	0.28	61	0.24	***
	Millet	0.04	26	0.06	18			0.28	29	0.20	20			0.50	36	0.52	25		
	Groundnuts	0.28	97	0.15	54	0.13	*	0.64	92	0.60	45			0.73	138	0.59	54		
	Yams							0.00	1	0.00	1			0.00	6	0.00	1		
	Cocoyams							0.00	1	0.00	1			0.00	1				
	Arrowroot													0.00	2	0.00	3		
	Vegetables	0.31	155	0.21	96	0.10	*	0.68	79	0.62	63			0.45	182	0.28	82	0.17	**
	Fruits							0.33	9	0.50	8			0.31	71	0.31	32		
	Other	0.22	23	0.10	10			0.61	128	0.56	91			0.84	56	0.79	14		
Tanzania	Cassava	0.50	68	0.30	10			0.37	35	0.38	8			0.28	102	0.15	20		
	Plantains	0.41	169	0.19	37	0.22	**	0.48	83	0.25	12			0.35	141	0.27	30		
	Beans	0.48	143	0.16	45	0.32	***	0.58	93	0.50	20			0.52	129	0.26	42	0.26	***
	Peas	0.34	145	0.33	30			0.39	57	0.83	6	−0.44	**	0.58	62	0.61	18		
	Irish potatoes	0.53	89	0.32	28	0.21	*	0.69	61	0.55	20			0.59	87	0.57	35		
	Sweet potatoes	0.28	197	0.24	51			0.38	55	0.31	13			0.32	131	0.23	43		
	Millet	0.44	63	0.36	14			0.31	16	0.33	6			0.00	1				
	Groundnuts	0.37	68	0.58	12			0.47	32	0.00	3			0.45	47	0.36	11		
	Yams							0.14	7	1.00	1	−0.86	*	0.17	12	0.40	5		
	Cocoyams							0.10	10	0.00	1			0.32	28	0.17	6		
	Arrowroot													0.00	3				
	Vegetables	0.37	219	0.29	58			0.68	85	0.59	17			0.46	184	0.49	55		
	Fruits							0.58	33	1.00	1			0.42	118	0.17	23	0.25	**
	Other	0.49	43	0.67	6			0.59	41	0.89	9	−0.30	*	0.62	56	0.60	20		

Zambia	Cassava	0.19	83	0.22	59			0.26	94	0.30	20			0.41	122	0.38	40		
	Plantains	0.40	47	0.75	4			0.54	41	0.50	4			0.45	109	0.29	24		
	Beans	0.30	115	0.17	41			0.38	117	0.33	24			0.44	142	0.36	44		
	Peas	0.67	3	0.25	4			0.46	13	0.75	4			0.00	7	0.00	5		
	Irish potatoes	0.50	10	0.00	2			0.48	21	0.50	4			0.65	20	0.67	3		
	Sweet potatoes	0.55	244	0.49	89			0.66	238	0.71	52			0.63	291	0.53	83		
	Millet	0.22	87	0.30	54			0.50	44	0.50	14			0.50	24	0.41	17		
	Groundnuts	0.41	209	0.31	80			0.53	246	0.60	55			0.63	304	0.53	83		
	Yams							0.00	3	0.00	2			0.50	2				
	Cocoyams									0.00	1			1.00	1				
	Arrowroot													0.00	1				
	Vegetables	0.74	163	0.66	47			0.87	134	0.88	25			0.73	231	0.73	49		
	Fruits							0.67	12	0.00	1			0.29	72	0.22	23		
	Other	0.74	47	0.58	12			0.33	78	0.31	13			0.65	85	0.59	17		
Mozambique	Cassava	0.32	180	0.23	126			0.27	201	0.30	50			0.44	176	0.31	88	0.13	**
	Plantains	0.42	64	0.40	29			0.46	57	0.36	11			0.47	102	0.44	36		
	Beans	0.24	127	0.18	104			0.11	164	0.08	72			0.42	166	0.24	118	0.18	***
	Peas	0.17	53	0.04	25	0.13	*	0.11	94	0.17	18			0.56	96	0.43	23		
	Irish potatoes	1.00	1	0.00	3			0.00	0	0.00	0			0.19	16	0.07	14		
	Sweet potatoes	0.12	85	0.10	68			0.17	52	0.25	16			0.37	134	0.17	100	0.20	***
	Millet	0.00	8	0.00	4			0.20	20	0.00	5			0.11	18	0.00	1		
	Groundnuts	0.54	134	0.30	94	0.24	***	0.54	136	0.43	30			0.56	165	0.24	95	0.32	***
	Yams							0.50	2	0.00	1			0.33	15	0.07	15	0.26	*
	Cocoyams							0.00		0.00	0			0.00	1	0.00	1		
	Arrowroot													0.29	7	0.00	3		
	Vegetables	0.13	85	0.11	66			0.42	24	0.13	8			0.41	49	0.18	51	0.23	**
	Fruits							0.83	6	0.80	5			0.40	50	0.33	18		
	Other	0.03	38	0.07	28			0.14	22	0.08	12			0.06	212	0.11	135		

Note: a. Data on yams, cocoyams, and fruits were added to the survey in 2008, arrowroot in 2013.

gender-based differences in the commercialization of maize, there were no statistically significant differences in market participation for any of the food crops in any of the years. The largest number of gender gaps in market participation is found in Mozambique, breaking with the trends for Afrint I and II.

In terms of marketing arrangements, participation in contract farming for other food crops has been very limited throughout the period from Afrint I onwards, both for MMFs and FMFs. Only in Ghana, in the final survey round, was the share of male farm managers participating in contract farming (4 per cent) higher than for their female counterparts, among whom none were engaged in contract farming (significant at the 5 per cent level).

What the data show is the need to address commercialization as a fluid process—farmers move in and out of markets, as opportunities change often in gender-neutral ways. Although the data do not detail amounts sold, the limited evidence of male domination of markets as food crops are increasingly commercialized (as seen in Tanzania) is counterbalanced by tendencies of increasing female biases in market participation in Kenya. Some country-level dynamics can be noted, however: Ghana and Zambia have few gender gaps in market participation for any of the non-grain food crops, while Mozambique and to some extent Malawi exhibit a growing masculinization of market participation for numerous crops. In the latter cases, moreover, this adds to male biases in maize commercialization found in these countries.

Non-Food Cash Crops

With respect to cash crops, again the patterns of production differ from country to country: although relatively few cash crops are grown in each country, the type of crop and the gendered patterns of commercialization vary. Despite such variation, the limited production of cash crops across the countries is quite striking and connected in large part to the regional nature of particular production systems. In Ghana only half the sampled villages are found in the cocoa-producing regions, for instance, explaining the low country-level involvement in cocoa production, despite its regional importance. With the possible exception of Kenya, where sugar cane and coffee are grown by around a quarter and a third of the sampled farmers respectively, cash crop production hence is generally limited.

Despite the common perception in the literature (Hill and Vigneri 2014) that traditional cash crops tend to be dominated by male farmers, Table 4.4 suggests that there are only three examples of cash crops whose production over the period is predominantly engaged in by male farm managers: sugar cane in Mozambique and Malawi, tobacco in Malawi, and cotton in Zambia. In the case of tobacco in Malawi, this dominance is explained by membership rules in the tobacco marketing association that discriminate against women,

Table 4.4. Share of households producing cash crops, by country and sex of farm manager

		Afrint I						Afrint II						Afrint III					
		Male	N	Female	N	Diff.	Sig.	Male	N	Female	N	Diff.	Sig.	Male	N	Female	N	Diff.	Sig.
Ghana	Cotton	0.01	342	0.00	69						111						113		
	Sugar cane	0.01	342	0.00	69						111						113		
	Cashew nuts	0.01	342	0.00	69			0.00	457	0.01	111						113		
	Cocoa	0.04	342	0.06	69			0.10	456	0.08	111			0.14	425	0.08	113		
	Tobacco	0.06	342	0.01	69			0.02	457	0.02	111			0.01	425	0.00	113		
	Coffee	0.02	342	0.00	69			0.01	457	0.01	111			0.00	425	0.00	113		
	Oil palm													0.11	425	0.03	113	0.08	***
	Other	0.05	316	0.03	64			0.01	457	0.01	111						113		
Kenya	Sugar cane	0.24	169	0.20	128			0.22	196	0.18	103			0.32	193	0.22	107	0.10	*
	Tobacco	0.01	169	0.00	128												107		
	Coffee	0.20	169	0.25	128			0.18	196	0.21	103			0.19	193	0.22	107		
	Tea	0.12	170	0.13	128			0.12	196	0.16	103			0.13	193	0.09	107		
	Sisal	0.01	169	0.00	128												107		
	Pyrethrum	0.01	169	0.01	128												107		
	Fruits and vegetables							0.11	196	0.07	103								
	Flowers							0.02	196	0.02	103			0.02	193	0.01	107		
	Spices							0.00	196	0.00	103			0.01	193	0.03	107		
	Other	0.08	167	0.05	128			0.02	194	0.01	102						107		
Malawi	Cotton	0.01	235	0.00	159			0.04	222	0.03	170			0.09	276	0.04	126	0.05	*
	Sugar cane	0.18	235	0.10	159	0.08	**	0.08	224	0.02	169	0.06	**	0.15	276	0.06	126	0.09	***
	Cocoa	0.02	235	0.03	159												126		
	Tobacco	0.17	235	0.09	159	0.08	**	0.15	224	0.02	170	0.13	***	0.14	276	0.02	126	0.12	***
	Coffee							0.00	224	0.01	170						126		
	Sisal							0.00	224	0.01	170			0.02	276	0.02	126		
	Pyrethrum	0.31	235	0.15	159	0.16	***										126		
	Other	0.00	233	0.01	157												126		
Tanzania	Cotton	0.01	322	0.00				0.00	318	0.01	74	−0.01	**	0.00	300	0.00	95		
	Sugar cane	0.04	322	0.00	81	0.04	*	0.05	318	0.00	74	0.05	**	0.08	300	0.08	95		
	Cashew nuts	0.01	322	0.00							74			0.12	300	0.05	95	0.07	*
	Cocoa	0.02	322	0.00				0.02	318	0.00	74			0.11	300	0.03	95	0.08	**

(continued)

Table 4.4. Continued

		Afrint I						Afrint II						Afrint III					
		Male	N	Female	N	Diff.	Sig.	Male	N	Female	N	Diff.	Sig.	Male	N	Female	N	Diff.	Sig.
	Coffee	0.05	322	0.00	81	0.05	**				74			0.00	300	0.00	95		
	Tea	0.01	322	0.01	81						74			0.01	300	0.00	95		
	Pyrethrum	0.02	322	0.01	81									0.01	300	0.02	95		
	Fruits and vegetables							0.01	319	0.00	74								
	Oil palm										74			0.13	300	0.06	95	0.07	*
	Flowers										74			0.01	300	0.00	95		
	Spices										74			0.02	300	0.00	95		
	Other	0.24	322	0.20	81			0.11	320	0.11	74			0.19	300	0.08	95	0.11	**
Zambia	Cotton	0.35	300	0.16	102	0.19	***	0.23	326	0.11	76	0.12	**	0.30	366	0.11	110	0.19	***
	Sugar cane	0.01	299	0.00	102			0.01	325	0.00	76			0.04	366	0.00	110	0.04	**
	Cashew nuts							0.01	326	0.00	76						110		
	Tobacco	0.00	299	0.01	102	−0.01	*				76						110		
	Other	0.11	299	0.01	102	0.10	***	0.12	326	0.09	76			0.01	366	0.00	110		
Mozambique	Cotton	0.02	227	0.00	170	0.02	*				121			0.02	242	0.00	183	0.02	*
	Sugar cane	0.14	227	0.06	170	0.08	***	0.11	281	0.02	121	0.09	***	0.25	242	0.13	183	0.12	***
	Cashew nuts	0.25	227	0.19	170			0.20	281	0.05	121	0.15	***	0.31	242	0.13	183	0.18	***
	Cocoa	0.01	227	0.02	170									0.01	242	0.01	183		
	Tobacco	0.03	227	0.01	170	0.02	*	0.01	281	0.00	121			0.02	242	0.00	183	0.02	**
	Tea													0.01	242	0.03	183	−0.02	*
	Pyrethrum													0.00	242	0.01	183		
	Fruits and vegetables							0.01	281	0.00	121								
	Oil palm													0.06	242	0.03	183		
	Spices							0.00	281	0.01	121						183		
	Other	0.01	227	0.01	170			0.06	281	0.02	121	0.04	**	0.05	242	0.03	183		

Notes: a. For Afrint I, there were many households that were coded as 'do not know' for cash crop production; it was assumed that these questions had been skipped rather than answered negatively, and these answers were recoded as zero.
b. 'Fruits and vegetables' refers to fruits and vegetables produced for export only. This item was added for Afrint II and removed for Afrint III.
c. Flowers, spices, and sisal were added for Afrint II.
d. Oil palm was added for Afrint III.

while travel to centralized auction floors is impossible for many women whose mobility is restricted by domestic responsibilities and gender norms concerning travel.

Cotton production in Zambia is contract based. Representatives of the two contracting companies, Dunavant and Alliance Cotton, as well as two lead farmers engaged in recruiting farmers by Dunavant in Nikantaka village, identified labour constraints, rather than marketing arrangements, as the major limit to women's production. Both contract schemes reported an increase in women growing cotton, although formal registration with the schemes may be carried out through adult sons for instance. Out of the ten lead farmers recruited by Dunavant in Chilekwa village to encourage cotton contracting, three were women when the qualitative interviews were carried out in late 2012. The quantitative data confirm widespread involvement in contract farming for cash crops in Zambia: for Afrint III, 82 per cent of the farmers who were involved in cash crop production grew cash crops on a contract basis, with no statistically significant differences based on the sex of farm manager. The major discrepancy between the two types of farms lies instead in production: whereas 118 male farm managers were involved in cash crop production only twelve female landholders were growing cash crops.

While long-term trends are visible only for three particular crops, at the country level the number of gender gaps in cash crop production are highest for Malawi and Mozambique, aggravating the biases found against FMFs with respect to maize commercialization as well as non-grain food crops in these countries. In the case of Zambia, gender biases in maize commercialization combine with poorer marketing possibilities for female farm managers with respect to cotton. To the extent that sources of gender gaps in commercialization can be identified, the constraints on maize marketing lies in the inability of female farm managers to produce a marketable surplus that competes with the volumes sold by male respondents, while male biases in cash crops appear to be connected to gender-segmented markets in the case of tobacco and to labour-related production constraints in the case of cotton. For Ghana and Kenya, again, few gender-based biases exist with respect to cash crop production, while for Tanzania the period between Afrint II and III has seen the emergence of a growing number of gaps between male and female farm managers.

Cash Incomes, Gender, and Diversification

Although the data on commercialization can tell us something about market participation they only provide a partial picture of livelihoods. Especially for those crops where data on volumes sold and prices received are lacking, information on cash incomes is a necessary complement to analysing the effects of market participation on livelihoods. Given the widespread and persistent bias against women in smallholder agriculture noted in the literature, the expectation would be that the combination of poorer assets and smaller commercial opportunities for female landholders would translate into lower cash incomes, especially from farm-based sources.

For Afrint II, when the size and age composition of the household are controlled for, there were statistically significant gender-based differences in cash incomes for two countries only: Malawi and Zambia.[1] As suggested by the data in Table 4.5, these income gaps persisted into the final survey round, with gaps emerging also in Ghana and Kenya. In Kenya, with the largest gap, members of female-headed households on average had cash incomes equivalent to 53 per cent of their male counterparts, compared with 65 per cent (Malawi) and 67 per cent (Ghana and Zambia) for the other countries.

Although male-headed households were advantaged in Zambia with respect to agricultural commercialization, especially in maize, in Kenya by contrast there were no statistically significant gender-based differences with respect to market participation in any of the crop types, except for sugar cane for which there was a very weak statistical significance. In the case of Kenya, however, the tenuous link between the data on commercialization and incomes can be explained by the fact that with the exception of maize we do not have information on the amounts sold for the particular crops. Similarly, the male biases in market participation for other food crops and cash crops are not reflected in the cash income data for Mozambique.

The farm-based sources of gender gaps in cash income in Ghana, Kenya, Malawi, and Zambia can be found by breaking the cash income per adult equivalent by the particular sources of income, as is done in Table 4.6. In the case of Malawi, sales of all types of crops are biased towards male farm managers in both years. For Zambia, the gender gap in maize sales is reflected in the income data on staple sales. Higher male incomes from cash crop sales are to be expected in Malawi and Zambia where production itself is segmented by gender. In Kenya and Ghana sales of animal products have become biased towards male-headed households, although for Ghana the largest gender gap is found in incomes generated from the sale of other food crops. On the whole therefore, countrywide trends in market participation and production of cash crops are also reflected in the various income streams.

Juxtaposing farm-based sources of cash income with incomes from non-agricultural activities points to some interesting differences between the households of male and female farm managers. In Afrint II, although

[1] This contradicts data presented in an earlier paper (Andersson Djurfeldt et al. 2013). The reason for the discrepancy is the treatment of extreme cases, which differs. In this chapter, extreme cases have been removed only for the final variable.

Table 4.5. Cash income per adult equivalent (mean and median), by country and sex of farm manager for Afrint II and Afrint III, in 2010 PPP-adjusted USD for households that earned cash income

	Afrint II								Afrint III							
	Mean						Median		Mean						Median	
	Male	N	Female	N	Diff.	Sig.	Male	Female	Male	N	Female	N	Diff.	Sig.	Male	Female
Ghana	316	441	327	107			136	159	445	394	297	107	148	**	161	148
Kenya	768	169	943	89			376	406	661	184	351	103	310	***	351	178
Malawi	144	214	114	162	29	**	102	61	94	257	61	115	33	***	64	34
Tanzania	286	272	273	61			180	105	285	276	249	88			180	142
Zambia	202	286	151	71	52	**	147	87	256	331	171	95	86	***	154	135
Mozambique	92	246	82	89			34	38	190	211	238	146			90	88

Notes: a. Exchange rates, consumer price index, and PPP index used refer to 2015 for Tanzania and Mozambique, for the remaining countries 2013.
b. Data include only households that had a cash income; extreme cases have been removed for the final calculated variable at the village level.
c. For 2008, data on household members below age 15 and above 61 had many cases coded as uncodeable and missing; these were recoded as zero, which may underestimate the size of households.
d. For 2013, the data quality was much higher and the uncodeable and do not know entries for household members below age 15 and above 61 were kept.
e. The large differences between median and mean cash incomes for both household types suggests that incomes are highly skewed, regardless of the sex of head of household.

Table 4.6. Cash incomes per adult equivalent by sex of farm manager and income source, Afrint II and Afrint III, in 2010 PPP-adjusted USD for households who earned cash income

		Afrint II						Afrint III					
		Male	N	Female	N	Diff.	Sig.	Male	N	Female	N	Diff.	Sig.
Ghana	Staple sales	98	440	103	106			75	394	40	107	34	*
	Sale of other food crops	89	441	90	107			189	394	79	107	110	**
	Sale of non-food cash crops	8	440	5	107			20	394	11	107		
	Sale of animal produce	35	441	33	107			18	394	11	107	7	**
	Leasing out machinery	1	441	0	107			1	394	0	107		
	Agricultural labour	1	441	3	107			4	394	2	107		
	Non-farm salaried employment	27	441	7	107			22	394	15	107		
	Micro-business	26	441	45	107			90	394	81	107		
	Large-scale business	11	441	9	107			0	394	0	107		
	Rent, interest	0	441	0	107			2	394	0	107		
	Pensions	5	441	0	107			5	394	0	107	5	*
	Remittances	15	440	33	107	−19	**	19	394	57	107	−38	***
Kenya	Staple sales	20	169	27	89			61	184	46	103		
	Sale of other food crops	223	169	273	89			70	184	51	103		
	Sale of non-food cash crops	146	169	143	89			140	184	75	103	65	*
	Sale of animal produce	224	169	159	88			133	184	77	103	56	**
	Leasing out machinery	0	169	0	89			5	184	1	103		
	Agricultural labour	27	169	60	89			22	184	16	103		
	Non-farm salaried employment	89	169	192	89			129	184	25	103	104	**
	Micro-business	18	169	48	88			38	184	27	103		
	Large-scale business	1	169	7	89			6	184	0	103		
	Rent, interest	0	169	0	89			12	184	4	103		
	Pensions	7	169	15	89			22	184	3	103	19	*
	Remittances	12	169	21	89			23	184	25	103		
Malawi	Staple sales	27	210	15	156	12	***	18	257	7	115	11	***
	Sale of other food crops	33	214	16	161	18	***	28	257	16	115	12	***
	Sale of non-food cash crops	20	213	3	161	17	***	10	257	1	115	8	***
	Sale of animal produce	8	212	9	162			5	257	3	115		
	Leasing out machinery	2	213	1	161	1	*	1	257	0	115		

	Agricultural labour	6	214	5	162			3	257	7	115	−4	***
	Non-farm salaried employment	30	213	40	162			15	257	14	115		
	Micro-business	15	212	21	160			12	257	8	115		
	Large-scale business	0	213	0	161			0	257	0	115		
	Rent, interest	0	214	0	160			1	257	1	115		
	Pensions	0	213	0	162			0	257	0	115		
	Remittances	4	210	6	159			2	257	3	115		
Tanzania	Staple sales	84	272	50	61	34	*	105	276	111	88		
	Sale of other food crops	41	272	10	61	30	**	41	276	22	88	19	*
	Sale of non-food cash crops	30	272	8	61			34	276	21	88		
	Sale of animal produce	21	272	11	61			13	276	10	88		
	Leasing out machinery	1	272	0	61			4	275	0	88		
	Agricultural labour	4	272	9	61	−5	**	7	276	10	88		
	Non-farm salaried employment	29	272	106	61	−78	**	9	276	5	88		
	Micro-business	68	272	60	61			60	276	61	88		
	Large-scale business	0	272	0	61			3	275	0	88		
	Rent, interest	4	272	1	61			3	275	1	88		
	Pensions	1	272	0	61			1	276	0	88		
	Remittances	2	272	17	61	−15	***	5	276	8	88		
Zambia	Staple sales	68	286	19	71	49	***	138	331	71	95	68	***
	Sale of other food crops	33	286	21	71			31	331	22	95		
	Sale of non-food cash crops	22	286	13	71			14	331	5	95	9	**
	Sale of animal produce	14	286	9	71			22	331	12	95		
	Leasing out machinery	3	286	2	71			2	331	2	95		
	Agricultural labour	9	286	18	71			10	331	11	95		
	Non-farm salaried employment	16	286	25	71			12	331	3	95		
	Micro-business	27	286	23	71			18	331	30	95		
	Large-scale business	0	286	0	71			3	331	0	95		
	Rent, interest	0	286	2	71	−2	**	1	331	1	95		
	Pensions	1	286	1	71			0	331	0	95		
	Remittances	9	286	20	71	−10	*	4	331	14	95	−10	***
Mozambique	Staple sales	12	247	7	89			20	210	11	145	9	*
	Sale of other food crops	5	247	1	89	3	**	17	208	14	144		
	Sale of non-food cash crops	4	247	1	89			11	210	12	145		

(continued)

Table 4.6. Continued

	Afrint II						Afrint III					
	Male	N	Female	N	Diff.	Sig.	Male	N	Female	N	Diff.	Sig.
Sale of animal produce	2	247	3	89			9	211	18	146		
Leasing out machinery	0	247	1	89	−1	***	1	211	0	145		
Agricultural labour	1	247	8	88	−7	***	7	211	6	146		
Non-farm salaried employment	18	247	18	89			33	207	82	143	−49	**
Micro-business	44	247	19	88	25	*	79	205	77	140		
Large-scale business	0	247	0	89			3	211	1	146		
Rent, interest	0	247	0	89			3	211	0	146		
Pensions	7	247	7	89			5	211	6	145		
Remittances	5	247	17	89	−12	***	6	211	17	146	−11	**

Notes: a. Negative differences denote a bias towards female-headed households, positive differences a bias towards male-headed households.
b. No extreme cases were removed.

gender-based differences were few, they were all biasing female farm managers, with the exception of micro-business in Mozambique. Incomes raised from remittances were higher among female-headed households in all countries except Kenya and Malawi. In addition, non-farm salaried employment was much higher for female respondents in Tanzania: 106 PPP (purchasing power parity) adjusted 2010 USD (US dollars) per adult equivalent compared with 29 for male farm managers. By the final survey round, remittances were still predominantly a female source of income in Ghana, Zambia, and Mozambique, whereas a relatively large male bias in salaried employment in the case of Kenya had emerged, contributing to the overall gender gap in cash incomes noted for Kenya. In general, however, the gendered segmentation among non-farm activities noted in the literature, where women are confined to low-entry, low-gain activities is not apparent from the data.

Commercialization and Livelihoods by Type of Female-Headed Household

As demonstrated in Chapter 3, asset availability among *de facto* and *de jure* female-headed households vary, especially with respect to land. The demographic profile of *de facto* female-headed households resembles that of male-headed households more closely than that of *de jure* female-headed households. On this basis, the expectation would be that there are differences in patterns of commercialization between the two household types. The data from Afrint III show no differences in the volume of maize sold, however, and a significantly higher share of maize production sold for *de jure* female-headed households, who sold 42 per cent of their harvest compared with 34 per cent for the *de facto* female-headed households (significant at the 5 per cent level). Mean output prices received for maize are the same for both household types.

For the other crop types, only two crops are grown by large enough numbers of respondents to enable comparison: groundnuts and vegetables. For these crops, commercialization is more pronounced among the *de facto* female-headed households. Groundnuts were sold by 42 per cent of the latter, compared with 22 per cent for the *de jure* female-headed households, while 58 per cent of the *de facto* female-headed households sold vegetables, compared with 44 per cent of the *de jure* female-headed households. The concentration of particular crops to the regions where the *de facto* female-headed households are strongly overrepresented (Nyeri and Kakamega in Kenya and the Centre in Mozambique) may explain these tendencies, however.

The *de facto* female-headed households had considerably higher mean cash incomes per adult equivalent than their *de jure* counterparts: 313 USD and 207 USD (PPP adjusted, at 2010 values) respectively. On average the incomes of the latter were more than 100 USD lower. While access to the incomes from absent husbands may be one explanation for this difference, another may be related to the concentration of *de facto* female-headed households to the wealthier countries in the sample (especially Kenya) and the relatively large share of *de jure* female-headed households in Malawi.

In terms of particular sources of income, two farm-based sources bias *de facto* female-headed households: sale of animal products and cash crops (both of which are important components of livelihoods in Nyeri and Kakamega). For the NFI streams, salaries and micro-businesses are also skewed towards these households, whereas somewhat counter-intuitively remittances favour *de jure* female-headed households. This hints at an age component, with remittances being likely to be related to transfers from adult children towards widowed women.

Distributional Aspects of Commercialization

The data presented so far show the existence of gender-based differences in commercialization in certain countries at particular points in time. In addition, they also suggests some variation with respect to commercialization between *de jure* and *de facto* female-headed households. Further variation may also be found *within* the broad groups of male and female-headed households. While female farm managers generally cultivate smaller areas, and also have lower access to other agricultural assets compared to male landholders (as demonstrated in Chapter 3), considerable heterogeneity may exist within these groups in terms of resources that affect productivity as well as opportunities for market engagement. As noted initially, maize is the most commonly grown and marketed crop across the six countries. Scrutinizing the patterns of maize commercialization with respect to socio-economic difference within the broad groups of MMFs and FMFs can provide insights into the variability of market engagement within these categories.

Housing standard constitutes a robust variable for tracing socio-economic differences across time and space—this metric is not sensitive to differences in cropping systems nor to fluctuations in exchange rates or PPP. While housing standards may be generally higher or lower in certain countries, the respondents with the highest housing standards are likely to be the best off in relative terms.

Three standards of housing were identified in the survey: mud house with thatched roof, mud houses with corrugated iron roof, and block/brick houses with corrugated iron roof, or other advanced housing. Here country differences are pronounced: by the time of Afrint III, in Tanzania with the highest overall housing standard around 60 per cent of the sample lived in the superior housing type, whereas in Mozambique just over a fifth did.

The ideal would have been to break the data down by the three types of housing standard, sex of farm manager, and country, but unfortunately the sample sizes are too small to enable this, so households with the highest housing standard are compared with the rest of the sample for male and female respondents respectively.

Some caution is needed when interpreting the results in Table 4.7: the number of cases is very small for the commercialization indicators that are based on production—especially for the female farm managers. This may explain why statistical significance is not attained even though differences with respect to commercialization are at times quite large between households of female farm managers living in advanced and basic housing respectively.

While this is clearly a rough measurement of wealth and welfare, some interesting results do emerge: for countries where market participation generally is high (Ghana, Zambia, and to some extent Tanzania), among the male farm managers commercialization is tilted towards respondents living in advanced housing. While there are few statistically significant differences for female-headed households, this distributional profile is not replicated, with the exception of volumes sold in Zambia, where differences are substantial between the better-off female farm managers and the rest. In the case of Zambia—where maize is most strongly commercialized, gender-based differences in commercialization are the most pronounced, as are the differences within the two groups. Overall, commercialization in Zambia is therefore biased towards male farm managers and towards the wealthier farm managers within the two household groups.

Intra-Household Income Generation

The results presented so far have concerned gender-based differences in commercialization and cash income sources between male- and female-headed households. While this to date is the most common method of analysing gender in studies of agriculture (Doss 2014), most women in rural Africa live in households headed by men.

The dataset does not cover individual data on production and commercialization, but it contains individualized cash income data by sex for Afrint III. Although these data are clearly a fractional representation of intra-household gender relations, they do enable identifying livelihood sources by sex, and intra-household complementarities in this respect (Jackson 2007).

Figure 4.1 shows the share of cash income raised by male members in the households of male farm managers. Here shares higher than 50 per cent point

Table 4.7. Commercialization in maize by country, sex of farm manager, and housing standard, Afrint III, for households who produced maize

		Male						Female					
		Basic	N	Advanced	N	Diff.	Sig.	Basic	N	Advanced	N	Diff.	
Ghana	Market participation	0.51	252	0.80	85	0.29	***	0.50	56	0.64	25		
	Share sold of production	0.52	129	0.65	68	0.13	***	0.51	28	0.59	16		
	Sale of maize (kg)	601	126	765	61	164	**	446	28	375	16		
Kenya	Market participation	0.40	85	0.29	108			0.29	52	0.38	55		
	Share sold of production	0.40	34	0.47	31			0.35	15	0.41	21		
	Sale of maize (kg)	727	33	643	30			357	15	503	20		
Malawi	Market participation	0.40	197	0.49	71			0.28	103	0.19	21		
	Share sold of production	0.20	78	0.25	35	0.05	*	0.22	29	0.30	4		
	Sale of maize (kg)	251	76	312	27			204	29	433	3		
Tanzania	Market participation	0.42	96	0.65	175	0.23	***	0.54	41	0.67	39		
	Share sold of production	0.41	40	0.44	113			0.50	21	0.49	26		
	Sale of maize (kg)	834	40	910	109			761	20	643	26		
Zambia	Market participation	0.75	208	0.86	147	0.11	**	0.57	82	0.68	25		
	Share sold of production	0.53	157	0.60	127	0.07	**	0.46	47	0.47	17		
	Sale of maize (kg)	2,080	155	4,165	114	2,085	***	974	46	1,721	17	747	**
Mozambique	Market participation	0.20	147	0.42	26	0.22	**	0.25	96	0.30	44		
	Share sold of production	0.26	28	0.46	11	0.20	**	0.25	23	0.19	13		
	Sale of maize (kg)	202	28	230	6			127	22	291	13	163	**

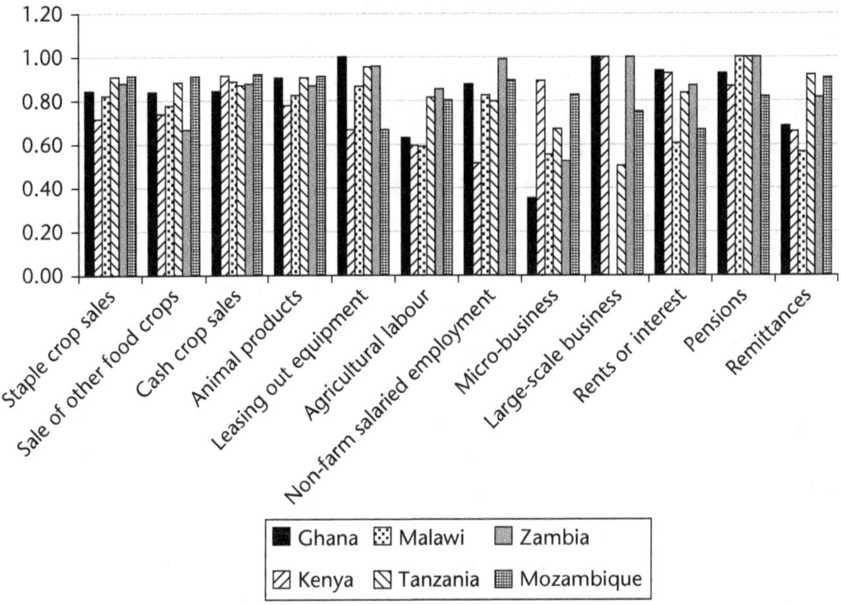

Figure 4.1 Share of cash income raised by male members in male-headed households, by income source and country, 2013.

to a male bias in particular income sources and those below 50 per cent indicate a female bias. This assumes on average an equal number of male and female members in the households of male farm managers, but given the large number of households, a systematic bias is not likely.

All farm-based sources of cash income generate higher incomes for men across the countries, pointing to a higher engagement with markets among men in male-headed households. In turn this suggests an intra-household division of labour where women are engaged mostly in unremunerated work—whether in production or reproduction. Qualitative data from Malawi confirm this division of labour—while agricultural production is carried out by both spouses, often with the explicitly stated aim of feeding the family, marketing decisions and the practicalities of commercialization, such as reaching markets and selling products, were generally the responsibility of the husband. The reasons for this were varied but revolved largely around the limitations to female mobility imposed by domestic responsibilities, but also in some cases the distrust expressed by husbands concerning their wives' ability to handle incomes from sales.

For non-farm activities, the differentiation of highly remunerative pursuits by gender referred to in the literature is only partially reflected in the data: in general incomes raised from non-farm activities are dominated by male household members, with a few exceptions. In Kenya, men and women raise income from non-farm salaried employment more or less to the same extent, while micro-business is biased towards women in Ghana. In Malawi and Zambia female and male household members generate roughly equal amounts of cash income from micro-business. On the whole though, the data show a prevalent bias against female members of the households of male farm managers, regardless of the type of income source.

Theoretical perspectives on female empowerment place great faith in the transformative capacity of women's income generation as a source of enhanced female bargaining power within households headed by men. Nonetheless, women across the three Malawian villages covered by the qualitative data all report that they are expected to disclose cash incomes generated both through the sale of agricultural products as well as non-farm activities to their husbands. Two women stated that they were required to do so but chose to defy this norm to preserve some financial independence from their husbands. While husbands in general also claimed to adhere to similar expectations from their wives, this was disputed by many of the female respondents. Moreover, even if cash income was used for family needs, in general one of the husband's prerogatives is to decide on the use of generated income, although this varies among villages under patrilineal and matrilineal tenure, with stronger female involvement in expenditure decisions in the latter (Andersson Djurfeldt et al. 2017 (resubmitted)).

Regional Perspectives on Livelihood Diversification and Gender

The data presented above have illustrated the contextual nature of agricultural commercialization and sources of income even at the country level. The point is often made, however, that gender relations as well as gendered patterns of production and marketing are context-specific, distinguished by regional or even village-level features. Indeed, the notion of women's crops is increasingly being questioned in the literature, and as shown in this chapter patterns of commercialization vary across time and space even at the macro scale.

To what extent differences in gender-based cash incomes can be attributed to regional production patterns and marketing arrangements that depart from and reinforce gender-based divisions of labour and commercialization patterns is the subject to which I turn next.

In Table 4.8, the sample is divided by region as well as by sex of farm manager, showing the concentration of gender-based income gaps documented at the country level in Table 4.5 to one region in Ghana and Zambia. In the case of Kenya and Malawi, while there are statistically significant differences for all regions, caution is warranted for those regions where statistical significance is only possible to establish at the 10 per cent level.

Table 4.8. Income per adult equivalent by sex of farm manager and region, Afrint III, in 2010 PPP-adjusted USD

		Male	N	Female	N	Diff.	Sig.
Ghana	Eastern	830	179	458	55	371	***
	Upper East	125	215	127	52		
Kenya	Kakamega	397	85	223	56	174	*
	Nyeri	888	99	504	47	385	***
Malawi	Ntchisi	82	76	52	18	30	*
	Thiwi Lifidzi	97	60	64	31	33	*
	Bwanje Valley	62	66	28	24	34	***
	Shire Highlands	146	55	81	42	65	**
Tanzania	Kilombero	327	152	278	36		
	Iringa	234	124	229	52		
Zambia	Mkushi	232	168	140	61	92	***
	Mazabuka	282	163	226	34		
Mozambique	North	90	119	84	30		
	Centre	356	76	365	80		
	South	142	15	84	36		

Moreover, differences in income are larger in one region in each country (Nyeri in Kenya, and Shire Highlands in Malawi).

Statistical testing of gender-based differences of means for particular sources of income is generally difficult, since small sample sizes result from dividing income data further. To the extent that statistical significance can be established, some interesting differences as well as similarities emerge: raising income through the sale of agricultural labour is biased towards the households of female farm managers in three regions (Nyeri, Kenya, and Ntchisi and Shire Highlands, both in Malawi)—but only in one of these is the difference statistically significant at below 5 per cent. Both regions in Zambia have higher incomes from staple sales among male-headed households, but only in the case of Mkushi does this translate into higher cash incomes overall. In the Eastern Region in Ghana, larger incomes raised from the sale of staple crops also explains higher incomes in male-headed households. In the case of the Eastern Region, incomes generated from the marketing of other food crops are also important. The vibrant tomato trade noted during fieldwork in 2011 may be a possible explanation for the income discrepancies between the two household groups (Andersson Djurfeldt et al. 2014). Depending on regional characteristics, other food crops and animal produce also generate more incomes for the households of male farm managers in some regions.

For NFI sources very few gender-based differences exist at the regional level. To the extent that they do, incomes from remittances are biased towards the households of female farm managers. Incomes raised from micro-business are also larger for female-headed households in Mazabuka, Zambia, and the two regions in Ghana. To some extent this compensates for lacking farm incomes.

Indeed, in the case of Mazabuka, Zambia, and the Upper East, Ghana, there are no statistically significant differences in cash incomes between male- and female-headed households. In the case of the Kenyan regions, by contrast, gender gaps in total cash income are related to non-farm sources of income. Salaried employment and pensions are skewed towards male-headed households in the case of Kakamega and Nyeri, respectively.

Although regional patterns of intra-household access to income within households headed by male farm managers largely replicate the country level, some interesting differences emerge between the regions. Three cases of farm incomes are biased towards women: the sale of other food crops in Mazabuka, Zambia, the leasing out of agricultural equipment in Kakamega, Kenya, and the sale of farm labour in Kakamega, Kenya and South, Mozambique. Leasing out of equipment is a very rare source of income, however, and the results should be treated with much caution.

Intra-household biases towards women are found with respect to micro-business involvement in four regions, Shire Highlands (Malawi), Eastern and Upper East (Ghana), and Mazabuka (Zambia). In the case of Mazabuka and the two regions in Ghana, this pattern mirrors the income gap based on sex of head of household, suggesting a stronger involvement of women in general in micro-business in these regions. Key informant interviews with the chief and headmen from Nikantaka village in Mazabuka points to the emergence of several non-farm opportunities connected to the Zambia Sugar processing plant in Mazabuka. Moreover, a nickel mine was established in the village in 2006 by an Australian company, providing jobs as well as demand for services. Although the mining company had pulled out of the village by the time of the interviews in 2012, female focus group discussions as well as interviews with traders in the local trading centre confirmed the continued importance and dynamism of small-scale trading and the engagement of women in this trade. A similar pattern was reported in the qualitative data for both regions in Ghana. In the Upper East there is a seasonal pattern to some of these activities, with artisanal small-scale mining and alcohol production and sale being carried out in the off-season. The villages in the Eastern Region are strategically located with respect to trade in vegetables as well as textiles with neighbouring countries.

In the two regions in Ghana and Mazabuka in Zambia, the segmentation of income opportunities by gender occurs both between male- and female-headed households as well as within male-headed households, but this stratification by gender appears to be complementary rather than competing. These regions are in the minority, however. Rather, the same patterns of lower female incomes characterize both the country and regional levels, pointing to the limited possibilities for women to raise individual incomes within male-headed household.

Conclusions

This chapter set out with the empiricist aspiration to address the gendered aspects of commercialization, diversification, and rural livelihoods. A number of key findings can be reported: agricultural commercialization, if measured through market participation, is not generally segmented by gender, although exceptions exist both nationally and regionally and with respect to particular crops. The data do not suggest a discrimination against female farmers with respect to prices received or the segmentation of particular marketing channels by gender, however. The exception here is tobacco in Malawi, where institutional discrimination against women prevents female participation.

The results also contradict the common perception in the literature that commercialization of food crops leads to male dominance of markets over time; rather, the patterns are fluid, with both male and female farmers moving in and out of markets as opportunities change. Nonetheless, particular countries stand out with respect to certain crops: for maize, a growing bias against female farm managers can be noted with respect to all measurements of commercialization in Zambia. Mozambique, Malawi, and to a lesser extent Tanzania stand out in terms of non-grain food crops, where market participation by male farm managers had increased relative to female-headed households between Afrint II and III.

In the case of maize, poorer commercial possibilities for female farm managers appear to be tied strongly to production factors, where lack of labour and land prevent the generation of a marketable surplus. The combination of domestic work with sole responsibility for farming in highly labour-intensive production systems was repeatedly mentioned in the qualitative interviews as the main difficulty facing female farm managers. While the gender-based gaps in total household cash income were limited to Malawi and Zambia during Afrint II, by the time of the third survey round, members of male-headed households were on average also earning higher cash incomes in Kenya and Ghana. The explanation for higher cash incomes within male-headed households can generally be found in higher farm-based incomes, with NFI sources earned by the households of female farm managers not being able to compensate for this relative shortfall. To the extent that agricultural commercialization has increased since Afrint II—as has been the case particularly in Zambia—it has benefited female farm managers but has also disproportionately advantaged their male counterparts.

The sources of intra-household gaps in cash income largely replicate the gaps found between male- and female-headed households, but intra-household gaps are in fact generally larger than the gaps found at household level. Male members generate the bulk of cash in households headed by male farm managers, whether from within or outside farming.

With respect to agricultural commercialization, the gendered patterns suggest an important distinction between women who manage their own farms and women who live in households headed by men. Whereas the constraints for the former appear to lie in the lack of access to agrarian resources that can be used to generate a marketable surplus, the disincentives to commercialization may be more forthcoming for the latter, as the outcomes from sales are controlled by their husbands. What unites both groups of women are the restrictions to commercialization imposed by lacking mobility, in turn related to domestic responsibilities and cultural mores that impede physical movement. Institutional segmentation of markets by gender, through male membership requirements in trading associations or contracting arrangements, may also exclude women.

Analysing the data by region rather than country shows that at this level income gaps are also explained by higher incomes from farm-based sources in male-headed households. By contrast, in some regions without gender-based income gaps, income generated from non-farm activities by female-headed households to some extent explains the absence of such gaps. While intra-household data at the regional level point to a handful of income sources (both farm- and non-farm-based) that are biased towards women, in general the data on cash incomes—whether at country, household, or individual level—point to the very limited possibilities for women to raise any type of cash income.

The mantra of 'linking smallholders to markets' has become something of an article of faith in pro-poor growth perspectives—with the 'business case' for addressing gender gaps in agriculture dovetailing with the 'social justice case' of enhancing women's empowerment through earning individual cash incomes. Theoretically female bargaining power can be improved through generating incomes either from agriculture or non-farm sources. The data suggest, however, that rural livelihoods are predominantly based on agrarian sources of income which in turn are based on productive assets that are unequally distributed by gender—whether among or within households. The prospects for linking female farmers to markets, therefore, are dimmed not by the institutional mechanisms of markets in themselves, but by the structural inability of female farmers to produce a marketable surplus. In this regard, non-farm sources of constraints on commercialization also need to be recognized, for instance the labour-intensive drudgery characteristic of women's lives in many parts of sub-Saharan Africa. While agricultural technology may be relevant to raising yields among female farmers through improved inputs for instance, non-agricultural interventions to ease constraints on women's labour and mobility, such as addressing the ill-health of young children and improving access to water for domestic use, may be equally important.

5

Agrarian Change and Structural Transformation: Drivers and Distributional Outcomes

Agnes Andersson Djurfeldt, Göran Djurfeldt, Ola Hall, and Maria Francisca Archila Bustos

Introduction

The structural transformation (ST) of developing economies is a much-studied process, but research is sparser on the topic of this chapter: the impact of ST on the agrarian economy and its structure, including gender segmentation. We identify three processes of agrarian change affected by ST: (1) *intensification* of grain cultivation, i.e. increasing per area yields of grains; (2) *commercial diversification of cropping*; and (3) increasing *non-farm diversification*, sometimes referred to as *pluriactivity* or *straddling* of sectors. Here the former term will be used when referring to livelihoods, which are pursued, hypothetically, to a growing extent by farmer households (Haggblade 2007, Haggblade et al. 2007). Two more processes—(4) increased use of agricultural labour and (5) increased rates of exit from the agrarian sector and entry into the non-agrarian one—are part of the ST but cannot be studied with the current dataset.

In a previous publication, two of the current authors studied the same processes for the first panel period (2002 to 2008); here we refer to this study as 'the previous study' (Andersson Djurfeldt and Djurfeldt 2013). In this chapter we deal with the second panel period (2008–13/15), studying households that were surveyed in both Afrint II and Afrint III. We ask if the process of ST-driven change is continuing and what its drivers and spatial and distributional profiles are. Although we see the processes of agrarian change that we study as driven by the ST, in the following we will be referring to the processes studied as agrarian ST or occasionally as transformation of the agrarian sector.

Agrarian Change Driven by Structural Transformation

The ST of a developing economy is technically defined as a process in which the non-agrarian share of the total GDP increases, which coincides with a decreasing share of the total labour force employed in agriculture. According to Lewis' classical theory, when in the early stages of ST the agrarian labour reserve has been emptied, ST depends upon an increasing productivity per agricultural worker (Lewis 1954). A global and longitudinal study by Timmer and Akkus shows that a global process of ST was going on from 1961 onwards, with a yearly decline of about 1 per cent in the global agricultural labour force (Timmer and Akkus 2008). Concurrently, global grain yields have increased (Neumann et al. 2010), as have rates of crop and non-farm diversification.

From the colonial era onwards, with the emptying of land reserves, agricultural development has been seen as a process of intensification propelled by population growth. The consensus in social science seems to be that population growth is not a driving force of intensification directly, except maybe indirectly and in a *longue durée* perspective (Hayami and Ruttan 1971, Platteau et al. 1998).

Increasing demand for staples from the non-agrarian sectors is stimulating commercialization of cropping and is an important, if not *the most* important driver of *intensification*. However, commercialization encourages the complementary process of *extensification* as well. The latter term refers to a process where total output is augmented by expanding areas under cultivation—with more or less constant yields. The balance between extensification and intensification is related to access to land and the growth of the farming population in a given area. The latter in turn is affected by the movement of labour from the agricultural sector, either through non-farm diversification, or through a complete exit from agriculture. Thus, both increasing demand for agricultural products, as well as increasing labour scarcity, condition the balance between extensification and intensification, and when the ST proceeds this is clearly in favour of the latter.

Diversification of cropping likewise is powered by the ST: as non-agrarian as well as agrarian GDP per capita increases, the demand for non-staple crops tends to surge, stimulating the development of markets for such crops. Diversification again may depend on new areas brought into cultivation (i.e. extensification), as well as on increased productivity (in economic terms) on already cultivated land (*intensification*). In sub-Saharan Africa at present, crop diversification often draws on micro-irrigation: in riverbeds, or in swampy or marshy lands. Traditionally such tough-to-till lands were not attractive to farmers, but with increasing demand they tend to be increasingly drawn upon (for an argument similar to ours see Adams and Carter 1987).

After this introductory section, we proceed with an overview of previous research before describing the data used and how they were produced. Then follows a discussion of hierarchical or multi-level models before expounding on the three models used. A descriptive analysis of the variables involved and the hypotheses associated with these is then made before coming to the results of modelling the process of *intensification, commercial crop diversification,* and *non-farm diversification.* A conclusion regarding the main hypotheses ends the chapter.

Previous Research

There are few studies focusing on agrarian ST, especially in sub-Saharan Africa. Most research is either on entire economies and their ST, or targeting the transformation of the industrial or service sectors. Among the valuable exceptions belongs the review of research and policy issues by Binswanger-Mkhize et al. (2010) and Fuglie (2011, see also Poulton draft 2017). They show how the agrarian ST in sub-Saharan African was more or less stagnant from the 1970s onwards, but suggest an increasing trend since around the turn of the millennium. With the global food price crisis from 2008 onwards, market incentives for producers improved. The consequences for smallholders of the long agricultural recession which effectively ended with the millennium was reflected in our previous study (Andersson Djurfeldt and Djurfeldt 2013).

In the earlier article, we discussed the trends of intensification for two periods—from the year of household formation to Afrint I and for the first panel period respectively. For the first period, we used retrospective binary data on whether the household increased production and cultivated area respectively since the year of household formation, which allows us to assess intensification patterns. On this basis, we estimate that during the period between the year of household formation until 2002, the intensification rate in grains among the farmers sampled in Afrint I was 1.3 per cent per year, but that it increased to 4.4 per cent annually during the first panel period. If the models of agrarian growth in the works cited above prove correct, we would expect even higher rates in the second panel period.

By contrast, our previous study of commercial crop diversification, measured by the sale of non-staple food crops, shows that this type of diversification declined somewhat between Afrint I and Afrint II compared to the period leading up to the first round of data collection. This might have been a reflection of push-driven diversification pre-2002 and improved possibilities for grain production after that year. If the macro-analyses quoted above hold water, we would expect agricultural diversification prompted by distress to be replaced by pull-driven processes, stimulated by increasing demand for non-staple crops, in its turn induced by improved purchasing power by local as well as urban-based consumers.

Another surprising finding from our previous study is that non-farm diversification or sector-straddling decreased during the first panel period compared to the period before that,[1] which may have been a result of push-driven factors. During the long agricultural recession during the 1980s and 1990s, many smallholders were pushed into the non-farm sector, but with the shift in the early years of the 2000s, and especially after 2008, the odds turned so that farming became more attractive. The question remains whether accumulation motives have replaced distress as the primary basis for non-farm diversification. If so, we would expect a faster rate of non-farm diversification for the second panel period.

A related question concerns the distributional outcomes of agrarian change prompted by the ST. Is it a pro-poor or a pro-rich process? Since the issues started to be studied around the turn of the nineteenth into the twentieth century, many scholars have claimed that the development of markets for farm produce necessarily leads to increasing polarization.[2] We discuss the distributional aspects of the three processes, including a specific focus on gender—a perspective that has been given limited attention in analyses of agrarian transformation.

Data Description

The Afrint surveys and panels are described in detail in Chapter 1. In this chapter we are drawing mainly on data from the second Panel 2, covering the survey rounds of Afrint II and Afrint III. Throughout we are comparing with Panel 1, data from which were used in the previous study (Andersson Djurfeldt and Djurfeldt 2013). Round 3 consists of 3,375 farm household cases, of which 2,138 are included in the panel. In the models presented here, 68 per cent of these are included and the rest are missing cases. Further details of the trends in the cross-sectional data can be found in Chapter 2.

It can be argued that attrition itself reflects the agrarian changes brought by the ST of the economy, for example in prompting people to migrate or leave farming. We have not attempted to model this, as Dzanku et al. have done for the Afrint Panel 1 (Dzanku et al. 2015).

[1] Here we cannot make an exact comparison with the pre-2002 period because comparable data were not collected.

[2] The development in Russia is a classic example where Lenin (1899, 1960) claimed that the development of capitalism led to expropriation of the peasants and the bifurcation of the peasantry into an agrarian bourgeoisie and a landless proletariat or poor peasantry, while famously Chayanov (1966; 1977) claimed the reverse, that the resilience of the peasantry was high, and that with the formation of a strong cooperative movement it could survive in a surrounding capitalist economy.

Hierarchical Modelling and Multi-Level Panel Survey Data

Strangely, the conventional way of analysing village survey data like ours often omit taking account of the complex sample structure, with units sampled in stages, from country, over region to village, and finally households or individuals. Given the complexity of calculating sampling errors in such designs (Kish 1965, Kish 1987, Cochran 2007) this is not surprising, although software for the purpose is extant and included, for example, in the 'complex sample module' in a package like IBM-SPSS. A convenient way of avoiding these complexities is to use hierarchical models, the design of which, at least in our case, mimics the design of a multistage sample. Specialized software has been developed for the purpose, for example MLWin, which is used here (Rasbash et al. 2010).[3]

A major advantage in using multi-level models is to avoid false positives, which are likely to arise, for example, in using village or country dummies in a multistage household sample such as ours. When doing so, every household gets a 0 or 1 value for the dummy, falsely implying that there are as many villages as households. When calculating the sample error, the result is an underestimation stemming from dividing by the root of the first-stage sample size, i.e. the number of households, instead of the root of the second-stage sample size, which equals the number of villages. One dare not guess how many false positives this has led to in development research where multistage household surveys are often used.

Modelling Techniques

Our response variables are all binary: (1) whether a farm household increased their yields of grain during the panel period, i.e. when we compare their yields during the 2006–8 period compared to their recent yields (2011–13); (2) whether they entered (or left) the market or increased their sales of non-staple food crops during the same period; and (3) whether they became non-farm diversified or ceased to be so during the panel period.

In the models for commercial diversification of cropping and for non-farm diversification we use the status at the end of the period (diversification and non-farm diversification dummy, respectively, for 2013) and then insert the same status at the start of the period (2008) as an autoregressive component. The core of the models then is the following:

[3] For a good introduction to these statistical techniques see Snijders and Bosker (1999).

$$y_{t1} = \beta_0 x_0 + \beta_1 y_{t0} + e_i \tag{5.1}$$

As formulated here the model is not yet multi-level: it shows how the dependent variable at the end of the period (y_{t1}) is regressed on the same variable at the start of the period (y_{t0}) and how (β_1) expresses the association between the two. Making the model autoregressive thus takes account of the fact that the status at (t_1) is dependent on that at (t_0). The extent of that dependence is caught in the regression coefficient (β_1) for the autoregressive variable (y_{t0}).

We define variables as multi-level with the subscripts (i), (j), and (k) referring to household (i), village (j), and region (k) respectively. Introducing this notation and two matrices of independent variables at household (i,j,k) and village level (j,k) yields the following type of model:

$$y_{t1ijk} = \beta_0 x_{0jk} + \beta_1 y_{t0ijk} + \beta_2 X_{1ijk} + \beta_3 X_{2jk} \tag{5.2}$$

The variance at household level is set to 1, while the village and region variance is modelled as:

$$\beta_0 x_{0jk} = \beta_0 + v_{0k} + u_{0jk} \tag{5.3}$$

where v_{0k} and u_{0jk} is the region and village-level variance respectively, so that the sum of the regression coefficient of the constant (β_0) and the two variance components make up the total residual (designated as (e_i in equation 5.1). Both variance components (u and v) in turn are defined as normally distributed with their means = 0 and their variances estimated by the sample variances.[4]

The *intensification* model needs a qualifying comment since the dependent variable is not a genuine dummy as are the ones for crop and non-farm diversification: you are either diversified in terms of cropping or income, or you are not. By contrast, the intensification process cannot be *a priori* defined as autoregressive: if a household intensifying during a period is not dependent on whether it did so during the preceding period. The variable refers to the *process* between t_0 and t_1, during which you either intensified or not. Thus we cannot use an autoregressive model for intensification as straightforwardly as we do for the other two processes. We chose instead to use the level of intensification of the village at t_0 as a pseudo-regressive variable, which proved to work fine.

The three models have been estimated, in a first stage by Iterative Generalised Least Squares (IGLS) and subsequently by Markov Chain Monte Carlo estimation using the IGLS parameter estimates as Bayesian priors. The list of independent variables refers to hypotheses about the drivers of the three

[4] More precisely as $v_{0k} \sim N(0, \Omega_v) : \Omega_v = \Omega^2_{v0}$ and analogously for the region level variance (u_{0jk}).

processes. While the exemplification above only covers three levels, in the application below we add country as the fourth level. Before considering the results of the modelling, we will look at the binary associations between the dependent variables and the vectors of independent ones. In connection with this we formulate the exact hypotheses to be tested in the models.

Descriptive Data and Hypotheses

As can be expected from a period of increasing pace of ST, all three processes studied seem to have accelerated for the households that were part of Panel 2 (cf. Table 5.1). The rate of intensification increased from 47 per cent in the first panel to 59 per cent in the second one. As can be inferred from the small standard errors, the difference between the periods is highly statistically significant. To be exactly comparable the mean estimates should be divided by the length of the panel in terms of years: we then get 6.71 per cent per year for the first period and 9.83 per cent for the later one.

For crop and non-farm diversification we compare the rates for the respective survey rounds: for Afrint I the rate of commercial crop diversification was 45 per cent; it had barely increased for Afrint II to 46 per cent. However, it took a leap between 2008 and 2013, jumping from 46 to 66 per cent. Percentage

Table 5.1. Rates of intensification and crop and non-farm diversification by panel period

a)	Panel period								
	Afrint I to Afrint II			Afrint II to Afrint III					
Rate of intensification	p	SE	n	p	SE	n			
per panel	0.47	0.006	1,601	0.59	0.007	1,321			
Mean % per year	6.71			9.83					
b)	Survey round								
	Afrint I			Afrint II			Afrint III		
Rate of diversification of cropping	p	SE	n	p	SE	n	p	SE	n
per panel	0.45	0.008	1,056	0.46	0.005	2,474	0.66	0.004	2,548
Mean % per year				0.14			3.33		
Non-farm diversification									
per panel	0.61	0.007	1,056	0.50	0.005	2,472	0.55	0.005	2,541
%				−1.57			0.83		

Note: a. p = proportion, SE = standard error, and n = number of cases.

Source: own panel data

wise we get a 0.14 per cent increase per year for the first panel, compared to 3.33 per cent for the most recent period. The acceleration would be expected during a process of high economic growth spurring the process of agrarian ST.

In Afrint I, 61 per cent of households had income from non-farm sources; this percentage dropped to 50 per cent by Afrint II (or a mean decrease of −1.57 per cent per year) and then rose to 55 per cent five years later (an increase of 0.83 per cent per year). A possible interpretation is that in the earlier period there was considerable push-driven diversification out of agriculture, a dynamic which diminished in the second period. The acceleration after that year could be evidence of pull- rather than push-driven non-farm diversification.

There is no doubt that the recent period of high growth for sub-Saharan Africa, stimulated by a period of high prices for raw materials in world markets (Fuglie 2011), impelled a growing pace of ST in the sample countries. This means an acceleration compared to the 2002–8 period, when the process of ST ran on a somewhat lower gear (compare the results in our previous study, Andersson Djurfeldt and Djurfeldt 2013). The current results should thus not be a surprise. The questions to be answered in the following relate to the drivers of these processes, and to the spatial and distributional consequences. Detailed descriptive statistics can be found in Table 5.2.

Independent Variable Selection and Formulation of Hypotheses

As we have already outlined, model design is based on matrices of independent variables at both household and village level, related to specific hypotheses about the drivers as well as consequences of the three processes. To counteract panel mortality we have replaced deceased respondents by sampling one of their descendants. Against a background of many studies showing that big changes in farming tend to occur as a consequence of the generational shift, we tested this hypothesis and the effect of increasing age in all three models. Including the latter variable also allows us to distinguish between the effects of ageing and time-bound effects more generally.

We will be looking specifically at FMFs. As can be seen from Table 5.3, 22 per cent of all sample farms were female-managed in 2013. We want to see if, when controlling for other factors, the odds are against FMFs when it comes to intensification and crop and non-farm diversification.

It is worth noting that, on the mean, household labour resources (measured as total number of family workers in agriculture) increased by 14 per cent[5] during the period, or on the average by 2.33 per cent per year. This is in line

[5] This is the antilog of the logged change reported in the table, 0.131.

Table 5.2. Descriptive statistics

Variable	Cross-section	Panel	Mean	SE
Autoregressive components[a]				
Village mean rate of grain intensification	Afrint II		0.310	0.025
		Afrint II–III	0.349	0.024
Other food crops sold	Afrint II		0.584	0.004
	Afrint III		0.662	0.009
Non-farm diversified household	Afrint II		0.575	0.010
	Afrint III		0.633	0.010
Age of head household, logged	Afrint III		3.883	0.006
Descendant household, dummy	Afrint III		0.050	0.004
Female-managed farm, dummy	Afrint III		0.220	0.008
Household labour resources logged, change in		Afrint II–III	0.131	0.016
Change in cultivated area		Afrint II–III	1.633	0.034
Extensifed grain production, dummy		Afrint II–III	0.572	0.011
Entered market or increased grain sales, dummy		Afrint II–III	0.380	0.011
No. of meals eaten in lean season	Afrint II		2.313	0.014
Access to household credit, dummy	Afrint II		0.110	0.006
Distance to nearest town with 50,000+ population, travel time	2000		2.680	0.042
Always used or started to use animal manure, dummy		Afrint II–III	0.872	0.009
Used improved seed since 2008, dummy	Afrint II		0.379	0.013
Belonged to 10% largest landholders in village, dummy	Afrint II		0.116	0.009
Land titling scheme present in village, dummy	Afrint II		0.500	0.069
Farm partly irrigated, dummy	Afrint II		0.386	0.049
Gini coefficient of land distribution in village	Afrint II		0.375	0.158
Diversified cropping since 2008		Afrint II–III	0.501	0.013
Became pluriactive since 2008		Afrint II–III	0.228	0.011
Intensified grain cultivation, 2008–13		Afrint II–III	0.301	0.012

Notes: a. Intensification is not a genuine binary outcome since the process of intensification can continue over time. The autoregressive component is replaced by a derived village-level mean (the share of farms in the village that intensified during the first panel period) in the intensification model.

with the opening of a demographic window of opportunity in sub-Saharan Africa, as shown by Masters (2011): the labour force is increasing in size and proportion both of the total as well as of the rural population. Increasing labour resources should, according to our hypothesis, favour intensification and diversification and cropping, and possibly also non-farm diversification.

Interestingly the average cultivated area per farm has increased by 63 per cent, or a mean of 9 per cent per year during the second panel. Much of this increase in cultivated area draws on fallows, according to our hypothesis, driven by increasing demand, both for grains and for other food crops. Expanded area is a condition for, as well as a consequence of, *extensification*, i.e. the increase in production by means of area expansion, rather than by higher yields. Fifty-seven per cent of the sample farmers have expanded their cultivated area since Afrint II, or an average of 9.5 per cent year. Our

Table 5.3. Models of three processes of agrarian structural transformation

Fixed effects	Model 1			Model 2			Model 3		
	Intensified grain cultivation since 2008			Sold other crops 2013			Non-farm diversified household 2013		
	β	SE	Sig.	β	SE	Sig.	β	SE	Sig.
Constant	3.33	0.77	***	1.46	1.07		2.17	0.93	**
Autoregressive component[b]	−1.56	0.59	***	0.99	0.15	***	0.75	0.14	***
Control variables									
Age of head household 2013, logged	−0.23	0.20		−0.09	0.20		−0.71	0.19	***
Head of household descendant of 2008 respondent	−0.29	0.31		−0.20	0.31		0.13	0.31	
Household characteristics									
Change in labour resources, 2008–13, logged	0.10	0.10		0.35	0.10	***	0.06	0.09	
No. of meals eaten in lean season, 2008	−0.31	0.11	***	0.17	0.12		0.01	0.11	
Belonged to 10% largest landholders in village, 2008	0.24	0.21		0.28	0.24		−0.55	0.20	***
Farm characteristics									
Female–managed farm	0.13	0.17		−0.78	0.16	***	0.15	0.15	
Change in cultivated area, 2008–13	−0.05	0.05		0.41	0.07	***	−0.14	0.05	***
Extensifed grain production, 2008–13	−1.13	0.16	***	−0.35	0.16	**	0.19	0.14	
Entered market or increased grain sales, 2008–13	1.59	0.16	***	0.22	0.15		−0.50	0.13	***
Always used or started to use animal manure, 2008–13	−0.10	0.25		0.13	0.21		0.22	0.19	
Used improved seed, 2008	−0.04	0.19		0.35	0.22		0.08	0.18	
Farm partly irrigated, 2013	−0.06	0.14		0.78	0.17	***	−0.25	0.14	*
Access to agricultural credit, 2008	0.02	0.22		0.65	0.29	**	0.12	0.21	
Village-level variables									
Distance to nearest town with 50,000+ population, travel time, 2000	−0.03	0.14		−0.61	0.15		0.24	0.14	*
Square of the above	0.00	0.01		0.11	0.02	***	−0.02	0.01	
Land titling scheme present in village, 2008	−0.61	0.27	**	0.69	0.28	**	0.34	0.25	
Gini coefficient of land distribution in village, 2008	−0.23	0.86		−4.34	0.78	***	1.34	0.59	**
Intercorrelation between dependent variables in Models 1 to 3									
Intensified grain cultivation since 2008	–	–	–	0.09	0.17		0.30	0.16	*
Diversified cropping since 2008	−0.09	0.15		–	–	–	0.02	0.15	

Became non-farm diversified since 2008	−0.01	0.16		−0.34	0.16	**	–	–	–
Intensified grain cultivation, 2008–13	–	–	–	0.09	0.17	0.17	0.30	0.16	*
Random effects[c]									
Country	0.13	0.34		0.13	0.39		0.45	0.31	
Region	1.20	0.62		1.48	0.80	*	0.82	0.25	**
Number of cases	1,447			1,447			1,447		
Missing (%)	32.32			32.32			32.32		
−2*loglikelihood, null model	2,005.31			1,673.53			1,853.86		
Ditto, full model	1,506.30			1,262.59			1,775.28		
Chi^2	499.0089	0.000	***	410.94	0.000	***	78.58	0.000	***
Degrees of freedom	25			25			25		

Notes: a. All regression coefficients have been tested. Significance levels are denoted as *, **, and *** for 10, 5, and 1 per cent respectively.

b. Intensification is not a genuine binary outcome since the process of intensification can continue over time. The autoregressive component is replaced by a derived villageylevel mean (the share of farms in the village that intensified during the first panel period) in the intensification model.

c. In a binary logistic model, the lowest-level variance (between villages) is set to 1. The level-specific variances are calculated as proportions of the former.

d. SE = standard error.

hypothesis is that although extensification is still very much a reality, compared to intensification it would be less important than it was for Panel 1.

Concurrently panel households have increased their herds of livestock by 0.48 tropical livestock units since Afrint II, or 8.0 per cent per year.[6] We would have preferred to use this variable to test a hypothesis that (so far) intensification is mainly dependent on sources of nutrients endogenous to the local economy and that exogenous sources, i.e. mainly inorganic fertilizer, are less important. Since livestock data on Mozambique are missing, using livestock ownership in the models implies that we lose one country and a large number of cases. However, data on the use of manure are available for all countries, and therefore we resort to these data in modelling.

The share of households that used manure rose from 60 to 87 per cent between Afrint I and II. However, usage fell again to 65 per cent by the final round of data collection. Our hypothesis is that, notwithstanding this recent decrease, intensification is mainly driven by endogenous sources of nutrients. Data on application of inorganic fertilizer are noisy, so use of manure is best compared with the usage of improved or hybrid seed, which was more or less constant for the first panel period (54 per cent compared to 52 per cent according to Jirström et al. 2011: 94). The percentage did not differ much for Afrint III, when it was 50 per cent (as reported in Chapter 2). In terms of crop technology then, we do not see much in terms of dynamism, so hypothetically crop technology exogenous to the village plays a minor role as a driver of grain intensification, as well as crop diversification. Given the recently boosted investments in diffusion of improved crop technologies and input subsidies across a range of country settings, this is a remarkable finding, which we return to later in the chapter.

More dynamism can be seen in the statistics on irrigation: While 32 per cent of the panel households had access to minor irrigation for the second round of data collection, the figure rose to 43 per cent by Afrint III, a yearly increase of 1.83 per cent. Note that this refers not to major irrigation systems, but to minor ones, constructed and managed either by communities or by individuals or groups of farmers. Diversification of cropping could be expected both to stimulate expansion of irrigation and to be stimulated by it; and intensification similarly, but less so since, of the grain crops studied, only paddy cultivation involves irrigation and moreover, most of the rice grown is rain-fed.

Afrint farmers face long travel times, on average two hours and forty minutes, to market outlets in bigger cities (more than 50,000 inhabitants). Travel

[6] Tropical livestock units (TLUs) are defined by a weighted sum of big as well as smaller livestock. Conversion factors are: cattle = 0.7, sheep = 0.1, goats = 0.1, pigs = 0.2, chicken = 0.01 (Harvest Choice 2011).

time and destination were estimated based on population figures for 2000 (Nelson 2008), so that presumably this is an overestimate compared to current conditions, which implies some noise in the data. Unfortunately, the dataset does not allow choosing a later period or smaller urban destinations, so these data serve as our indicator of market accessibility.[7]

Curve estimation exercises have been made for our three outcome variables (intensification and crop and non-farm diversification) regressed on travel time, with equations including the square of time. These show that there is a negative correlation between travel time and intensification, as well as with non-farm diversification; on the other hand we see a positive association with commercial crop diversification. The former correlations are expected: the farther to the major market, the lower the propensity for farmers to intensify or diversify into NFIs. In the multivariate setup that follows we use statistical controls to explore further the association between travel time to market and the three processes under investigation.

Note that the positive association between commercial crop diversification and travel time defies the expectation that the propensity for a farmer to diversify would decrease with increasing distance to the market. However, non-staple food crops are often sensitive to transport over long distances, especially under African conditions, therefore production of such crops is typically directed towards local markets, rather than regional, national, or international ones.[8] This could explain the positive correlation between distance to markets and commercial diversification within agriculture, and implies a modification of the overall hypothesis on the importance of distance to markets.[9]

There are many prerequisites for a deeper smallholder market integration of smallholders. Infrastructure is obviously important, but so is institutional development (Platteau et al. 1998). We use two indicators of such development: availability of agricultural credit and land titling. While for Afrint II, 12 per cent of the panel households had access to credit, the percentage had substantially increased by the third round of data collection when it reached 19 per cent, a difference which is statistically significant. We have no updated figures when it comes to changes in land tenure, but already for Afrint II, 47 per cent of the panel households were living in villages, which had land titling schemes. The question is to what extent these institutional changes are contributing to the agrarian change induced by the ST. According to our

[7] It should be pointed out that we have several survey indicators of distance, travel time, and cost of transport to market, but all these seem inferior to the one used here.

[8] Only one of the sample villages is involved in production directly for the international market.

[9] In all three cases the square term is statistically significant but of such a low magnitude that it hardly affects the estimate of the dependent variables (propensities to intensify, diversify, or go pluriactive).

hypothesis they will assume an increasing importance, especially for intensification and diversification of cropping.

Our models also deal with the distributional patterns of the agrarian ST, especially in terms of poverty and distribution of assets. The most reliable poverty indicator in our case seems to be the number of meals eaten in the lean season, which on the mean were 2.3 in Afrint II. One of the hypotheses that we test is whether intensification is a pro-poor process, and if this holds for non-farm diversification as well. Commercial crop diversification, on the other hand, could arguably be a strategy for farmers who are relatively food secure and possess the resources in terms of various forms of capital to venture into the more risky strategy of commercial diversification. The latter strategy then is not expected to be pro-poor, while the two first-mentioned ones could very well be.

We study only one indicator of wealth, namely landownership, and more specifically its distribution proxied by the Gini index. Varying between 0 and 1 the Gini is the most frequently used inequality measure; it was 0.38 for Afrint II. Our expectation is that agrarian ST, while not necessarily pro-poor, should not be anti-poor.

Another indicator of unequal distribution of benefits is if a household belonged to the 10 per cent of largest landowners in a village at the start of the period. Based on previous research on the Afrint data, we expect that the three processes we are studying do not advantage village elites. Thus, we expect a zero or negative correlation between the dependent variables and the elite indicator.

Comparing the results for the three variables mentioned, i.e. the poverty indicator, the Gini index, and the elite dummy, will enable us to assess whether the processes are more attractive to middling farmers compared to either the poor or the rich elite.

The final question is whether the three processes are interrelated. To test for this we include the change during the second panel for the respective response variables as independent variables in the other equations. Note, however, that even if at the household level the processes may be uncorrelated, this need not imply that they are so at the village level.

A final hypothesis deals with the autoregressive effect in all models, which we expect to be stronger than in the corresponding models in the preceding study (Andersson Djurfeldt and Djurfeldt 2013). The consequence of the accelerating ST process should be reflected in the coefficients for the autoregressive components, which should not only be positive, but higher than in the previous panel period. However, as already pointed out, only two of the models, i.e. those for crop and non-farm diversification, are genuinely autoregressive. The intensification model is different, with a 'pseudo-autoregressive' component, i.e. the village mean rate of intensification for the first panel period.

On the whole, using panel data considerably simplifies causal attribution: when working with cross-sectional data this is a perennial problem which statisticians and economists have shown considerable ingenuity in addressing. With panel data we always have the possibility of using lagged variables, or factors measured at t_0 rather than t_1. Since causality is not retroactive, the lagged variable may be a cause of the independent ones, but not the other way around. However, in this case it is not always possible to use the lagged variable strategy; in some cases, we use variables referring to changes in the panel period as a whole (t_0 to t_1) rather than to a state at t_0. Since the dependent variables refer to the same period, this makes causal attribution less straightforward. In most cases, however, we are interested in the association between the two variables; or differently put: what is the correlation between the variables, given that we control for other variables, as we do in multiple regressions? The cases where the dependent and independent variables refer to the same time period are discussed in detail in what follows.

Modelling Results

Table 5.3 presents the three models of (1) intensification, (2) commercial diversification of cropping, and (3) non-farm diversification.

All three models are highly statistically significant, with Chi^2 tests significant below the 0.1 per cent level. Regression coefficients are significant when marked with asterisks. For the crop diversification model, however, the constant is not statistically different from zero.

Model 1: Intensification

In the model of intensification, four regression coefficients are strongly statistically significant while one is significant at the 5 per cent level. First, the village rate of intensification for the first panel period is negatively related to the odds for a household in that village to have intensified its grain cultivation during the second panel. We interpret this as a saturation effect: in villages with high rates of intensification during the first panel period, the odds for an individual household to have intensified in the second period would be much lower than elsewhere. This is undeniably a positive distributional consequence of the process of ST as it is currently evolving since more villages, previously less intensified and presumably farther from the market, are apparently being drawn into the process.[10]

[10] This could have the effect of weakening the association between the odds for intensification and distance to market.

The indicator of the twin process to intensification, i.e. extensification, is also highly statistically significant, but negative ($\beta = -1.13$); while it was positive and highly significant in the earlier period ($\beta = 0.98$). Here, both the dependent and the independent variables refer to the same time period, namely the second panel. However, in this case we are interested in the association between them, rather than seeing one factor as the cause of the other. Thus we can assert that, from having been complementary options, intensification and extensification are becoming mutually exclusive strategies. For many farmers the option of increasing area, rather than yields, has been a relevant option: they may have done so, either at the relative expense of other crops, or absolutely by drawing on fallows or virgin land. In other words, the increase in grain production stimulated by increasing demand continues to take two, mutually exclusive forms: extensification *or* intensification.

The commercialization indicator strengthens this interpretation: farmers who have entered the grain market, or increased their volume of sales in that market, are almost five times ($Exp(1.59) = 4.89$) as likely to have increased their yields compared to those who decreased their sales or left the market during the period. Causal attribution is not self-evident in this case, since both the test variable and the dependent variable refer to the same time period. In principle, causation can be either from intensification to commercialization, the other way around, or possibly even circular. Whatever the case, there is no doubt about the strong association between the two variables. In the previous study the corresponding regression coefficient for the first panel period was of a similar order, but lower or ($\beta = 1.29$). If anything then, commercialization increased in importance in the later period. This reinforces the conclusion drawn in our earlier study and supports the hypothesis about an accelerating ST. Commercialization driven by increasing demand is not only the strongest factor associated with grain intensification (Andersson Djurfeldt and Djurfeldt 2013) but moreover a quickening one.

The spatial dimension of intensification should be possible to gauge by the distance indicator. On the binary level, as we have seen, intensification is negatively related to the travel time to the nearest major market. However, in a multivariate setup the regression coefficient is not statistically significant. This is a setback for our hypothesis, which however is likely to be an effect of other variables, collinear with distance, for example the commercialization indicator or the indicators of crop and non-farm diversification, respectively. Moreover, as already pointed out, the pseudo-autoregressive component, the village-level rate of intensification, could have a similar effect.[11]

[11] In a model stripped of a number of variables which are not significant in the one above, and with the use of manure replaced by the number of tropical livestock units, we get the expected

Since the turn of the century, African governments have given increasing attention to food production and security, with yield-increasing programmes receiving more funding from governments and donors.[12] Against this background it may be disappointing that the use of improved seed cannot be shown to have had any influence on the intensification process.

Increased yields might instead be conditioned by access to manure.[13] As is clear from Table 5.3, however, the proxy for the use of manure is not statistically significant. In a stripped model (see footnote 11) the use of manure is replaced by a less noisy variable: the number of tropical livestock units owned (details withheld).[14] The elasticity of intensification to a unit increase in livestock ownership in this model is ($\beta = 0.32$; 5 per cent significant), implying that adding a unit of livestock increases the odds for having intensified by 37 per cent. As far as can be judged from our data, intensification thus depends on local, organic sources of nutrients, which may limit attainable yields over time.

Credit and land titling may stimulate intensification according to our hypothesis: studying the influence we find that credit shows zero effect on intensification. As discussed later, this may suggest that insofar as intensification implies investment it may be financed through means other than credit, for example NFI. Land titling, on the other hand, proves negatively related to intensification but—as we will see—is positively related to commercial crop diversification. This will be further discussed in what follows.

Looking at other distributional consequences of intensification, we note that, like in the earlier period, elite membership, as proxied by belonging to the 10 per cent of largest landowners in a village at the start of the period, does not increase the odds of having intensified. The process does not appear to be pro-poor either, since the association with the poverty indicator[15] is negative and statistically significant: a household that had had one meal less than the mean number of meals taken during the lean season in 2008 had 27 per cent lower than average odds of having intensified its grain production since then ($Exp(-0.31) = 0.73$). A quite significant conclusion is thus that

significance for travel time ($\beta = -0.16$), significant at the 5 per cent level. This adds to the evidence that the odds for intensification increase with decreasing travel time to markets.

[12] Refer to the Comprehensive Africa Agriculture Development Programme under New Partnership for Africa's Development, which has meant increasing funding for agricultural development, both from governments and from donors (Kolavalli et al. 2010).

[13] Here both the dependent and the independent variable refer to the same time period, 2008–13. Again, the issue is not one of causal attribution but rather of association between the variables.

[14] Livestock ownership is measured in tropical livestock units, according to standards established by the United Nations Food and Agriculture Organization (Harvest Choice 2011). This information was not collected in Mozambique. For this reason, we do not include this country in the stripped model.

[15] The number of meals eaten in the lean season, 2008.

intensification (neither elite or pro-poor) seems to be a process attracting middling smallholders.

Regarding gender finally, intensification cannot be shown to discriminate against or privilege FMFs. Thus intensification is not obviously gendered.

The country- and regional-level variances are not statistically significant. This would strengthen the conclusion drawn in the earlier study, that there are no great differences in the intensification processes as they play out in different countries and regions. We seem to be witnessing a generic process.

Model 2: Commercial Crop Diversification

The autoregressive component in the diversification model (if the household was commercially diversified at the start of the period or had increased its market engagement) is strongly significant and with a regression coefficient ($\beta = 0.99$), as seen in Table 5.3. This implies that the odds for the household to be diversified during the second period are considerably higher than if the household had not been diversified in the previous period ($Exp(\beta) = 2.69$). Thus, once diversified, households are not disposed to withdraw from non-staple food crop markets, another sign of their being drawn into an ongoing process of ST.

Diversification is lower among female farm managers: the odds for such a household to have diversified is ($\beta = 0.78; Exp(\beta) = 0.46$), which is around 50 per cent less than for male farm managers. This implies that, in contrast to intensification, diversification seems to be a gendered process, tilted against female farm managers. This interpretation is supported also by the cross-sectional trends presented in Chapter 4 of this volume, which suggest a masculinization of markets for certain non-staple food crops during the second panel period for Mozambique, Malawi, and Tanzania.

Like extensification in grains, diversification of cropping to some extent draws on added landed resources: the relevant regression coefficient implies that the elasticity of diversification to an increase in cultivated area is ($\beta = 0.41;$ $Exp(\beta) = 1.51$). This is more than double the figure noted for the first period, again an indicator of progressing ST and in line with our hypothesis.

Commercial diversification of cropping is highly responsive to an increase in labour resources. T regression coefficient is ($\beta = 0.35; Exp(\beta) = 1.42$), meaning that the elasticity of commercial diversification to a 1 per cent increase in labour resources is 0.42 per cent. This once again underscores the importance of market incentives to smallholder commercialization.

Interestingly, in agreement with another of our hypotheses, diversification seems to be driven by access to credit, with a regression coefficient of ($\beta = 0.65; Exp(\beta) = 1.92$), indicating that a household with credit access had almost double the odds of crop diversification for the second panel period, compared to households without such access. In the previous period we found no indication of diversification facilitated by credit (Andersson Djurfeldt and Djurfeldt 2013). Furthermore, if land titling was present in a village at the beginning of the period, the odds for diversification are considerably higher. None of these indicators were significant in the previous period, pointing to changing conditions and contexts for the agrarian transformation process. Another sign of institutional development in other words.

Besides the further development of markets for staples and non-staples, this points to an institutional development, itself a part of ST, creating conditions for further agrarian transformation. Despite much criticism against land titling superimposed on customary land tenure systems, this result seems to suggest that land titling may fulfil its professed function of facilitating commercialization and reducing risks. However, it obviously says nothing about the possible social costs of such policies and gender aspects in particular. In this context, it can be noted that several authors note that formalization of customary tenure systems leads to weaker access and user rights among marginalized groups such as women and migrants (Walker 2003, Meinzen-Dick and Mwangi 2009, Razavi 2009, Alden Wily 2011).

Longer distance to markets would be expected to mean lower odds for diversification, but apparently this does not hold and is against expectations. The regression coefficient for travel time to major urban markets is not statistically significant, while its square is positive and significant. This points to a U-shaped distribution and indicates that when controlling for other factors, the odds for crop diversification first decrease and then increase with increasing distance. This partly contradicts the binary relation reported earlier, which was positive but not U-shaped. Furthermore, remembering the non-significant association between travel time and the odds for intensification in Model 1, it seems that multicollinearity may be a problem here too. A stripped model (see footnote 11) gives a positive regression coefficient, which corroborates the suspicion that multicollinearity is disguising what is probably a positive association. Thus we retain the interpretation that commercial crop diversification is often directed to local rather than regional markets.

Like for intensification, commercial diversification of cropping does not appear to be favouring local elites, while the regression coefficient for the poverty indicator likewise is not significant. Our model thus indicates that diversification of cropping is neither pro-elite, nor pro-poor, albeit slanted against women.

Investment of family labour and capital in irrigation is often a prerequisite for diversification. The process draws on thirsty crops like vegetables, moreover often grown in the off-season. The regression coefficient for access to (mainly minor) irrigation is 0.78, meaning that farms with access to irrigation have more than double the odds of being diversified compared to those without.

Furthermore, there is a strong association between inequality of landownership in the village, as measured by the village Gini index for land distribution ($\beta = -4.34; Exp(\beta) = 0.01$) and the odds for diversification of cropping. Thus the higher the inequality in landownership in a village, the lower the odds for crop diversification. To elaborate the conclusion already reached: crop diversification is neither pro-rich nor pro-poor, but seems more likely in village with many middling farmers.

While there is no intercorrelation between the odds for crop diversification and intensification, the process is negatively associated with non-farm diversification. At the household level the processes thus seem to some extent opposing, such that households tend to go either for one or the other.

As for the variance decomposition, there are some outliers at village and regional level but no clear ones at country level, reflected in a statistically significant region-level variance, but a non-significant one at country level. Diversification thus seems to be a uniform process, perhaps uneven in pace and space but with the same drivers.

Model 3: Non-Farm Diversification

In the study of the first panel period we had difficulties in designing a model of non-farm diversification that was at all statistically significant and thus had trouble in attaining statistical significance for the independent variables (Andersson Djurfeldt and Djurfeldt 2013). We did not realize that this was probably due to over-fitting and the large number of non-significant independent variables included in the model. Although it contains a fair number of independent variables the current model (cf. Table 5.3) is much more sturdy. The good fit probably depends on better data quality as well.

Age and gender give comparable results for both periods and in line with the hypothesis. The former variable is negatively correlated with non-farm diversification both for the first panel and for the second panel. The regression coefficient is considerably higher in the latter period though, with regression coefficients of $\beta = -0.20$ for the first period and -0.71 for the second period. If non-farm diversification were a *push*- rather than *pull*-driven process in the late twentieth and early twenty-first century, and if the ST implies that it is becoming less so, this would explain, at least ad hoc, the lower negative correlation between age logged and the odds for non-farm diversification. Increased food and livelihood security in the more recent period would, in this interpretation, be pushing fewer old people into low-return micro-business, like peddling vegetables to passers-by along the roadside.

The overall increase in labour resources that we have already noted is not associated with non-farm diversification while, recalling results for both

Models 1 and 2, it is positively associated with crop diversification and not at all with intensification. This may imply that, while all three strategies would be facilitated by increasing labour resources, crop diversification as a vent for 'surplus labour' is a more feasible option than the other two.

For the second panel, in contrast to the first one, we get a non-significant regression coefficient for the gender of farm manager. Unlike crop diversification where FMFs are less likely to be involved, non-farm diversification, like intensification, seems to have included both types of households. Thus the gender dynamics of the processes are mixed, at least at household level.

We get a weak statistical significance for the association between the odds for non-farm diversification and the distance indicator, but would have expected a stronger one. Recalling that the binary correlation was negative, we suspect another problem in this model compared to the previous ones. The outcome variable is heterogeneous, containing small-scale as well as large-scale businesses, casual employment as well as salaried work, alongside pensions and remittances. When we run a model regressing household involvement in micro-business on a stripped number of independent variables (cf. footnote 11), we do get the expected strong association with distance: the longer the travel time to markets, the lower the odds for being involved in micro-business. This suggests that the ST brings with it tighter and more frequent relations between urban and rural areas, manifested in increased non-farm diversification around the urban centres.

Model 3 reveals some other features of non-farm diversification, which deserve to be commented on. The autoregressive component is high and statistically significant ($\beta = 0.75; Exp(\beta) = 2.12$), which is expected and implies that once diversified into non-farm sources of income, households are highly likely to remain so. This contradicts data from our earlier study, which suggest mobility into and out of the non-farm sector in response to challenges as well as opportunities in the farm sector. In its turn, this contrast may point to the consolidation of non-farm diversification during the second period in response to broader structural change of the macro-level economy. 'Straddling' the farm and non-farm sectors, as it has also been called, should not be regarded as a temporary response to crisis, but as an enduring livelihood strategy, involving an increasing number of smallholders globally (Djurfeldt and Sircar 2016).

Furthermore, non-farm diversification is negatively related to an increased cultivated area, indicating that for many households this strategy may be an alternative to a more vigorous engagement in farming. The strongly negative association with grain market entry points in the same direction, again suggesting more permanent livelihood strategies focused on the non-farm sector.

Non-farm diversification is not significantly associated with the poverty indicator, but negatively so with elite membership, signalling that the strategy is widespread among other sections of smallholders. Not unexpectedly it is unrelated to credit access, as well as to the usage of manure or improved seed. Non-farm diversification is moreover negatively related to access to irrigation, but not associated with presence of land titling schemes in the village. Finally, the strategy is much more common in villages with unequal distributions of landownership. Rather than remain in agriculture or work as agricultural labour for neighbours with larger plots of land, it seems many smallholders have diversified into non-farm sources of income.

All this points to non-farm diversification to some extent being an alternative to commercial crop diversification, but not to intensification with which it is positively related ($\beta = 0.30; Exp(\beta) = 1.35$). Households engaged in non-farm diversification are thus more likely to intensify their cultivation than others. This piece of evidence obviously points to non-farm diversification being an alternative to credit for financing intensification. Note also that the strategy is unrelated to extensification, which may mean that non-farm diversification may lessen the pressure on land in a region.

Variance decomposition gives results similar to the previous models, with few deviant villages, but like in Model 2 with some regions significantly differing from the overall trends. Countries differ little and what we observe may therefore be a general trend.

Conclusions

On a speculative note, it is tempting to see the three periods covered in this chapter and in the previous work as pointing to three modalities in the relation between African smallholder agriculture and the non-agrarian economy, exemplified by: (1) a period with slow growth in the non-agrarian economy, creating sluggish demand for farm products and for what Arthur Lewis would have called surplus labour; (2) another period with more vigorous growth in the non-farm sector boosting the demand for staples as well as non-staples; and (3) yet another period with sustained non-agricultural growth and strengthened demand for farm products.

The first of these periods is easily recognized as the long recession in African agriculture from the 1980s until the early years of the twenty-first century (Fuglie 2011). The concurrent agrarian stagnation was reinforced by low world market prices, maintained by the dumping of farm surpluses by the OECD countries, tempting African governments to feed their urban populations by cheap imports rather than by domestic production.

Tables turned with the period of growth from around the time of the first survey, which was founded on high raw material prices stimulating the African non-farm sector, foreign direct investment, and GDP growth. During similar previous periods, growth occurred in spurts and collapsed with a fall in raw material prices, but from 2008 it was reinforced by the global food prices crisis, which brought about a period of higher world market food prices, and indirectly increased demand for domestically produced staples. This situation still prevailed when writing this in early 2017.

The models we have developed in this and our previous work illustrate the response of African smallholders to these different economic circumstances. By modelling household strategies from the year of household formation (*c.*1980) to the time of the second round of data collection (Afrint II) in the previous work (Andersson Djurfeldt and Djurfeldt 2013), we saw the effects of sluggish growth on staples, as well as non-staple production. Combined with population growth and increasing land scarcity, smallholder households were pushed into low-return off-farm activities, like micro-business and low-wage labour, on as well as off the farm. In the 2013 work mentioned above, we witnessed the turnaround from a low-growth scenario pre-2002, fed by sluggish demand for farm products, to one of higher growth and demand. As a reaction to this, smallholders withdrew from low-return non-farm activities and from less remunerative non-food cash crops. Stimulated by improved prices, they tended to intensify first of all in staple crop production, possibly with a view to their own food security. This tendency was further boosted during the second panel period when members of smallholder households increased their diversification, on as well as off the farm. This time it appears they were pulled into the non-farm sector by better prospects rather than pushed out of the farm sector.

In line with the above, the three indicators of ST that we are using here indicate an increasing pace of transformation (cf. Table 5.1). During the process of ST, one would also expect an increasing demand for labour, leading to improved wage levels for farm workers. This issue has not been researched here, but hypothetically sub-Saharan African agriculture, outside some pockets, has not yet come to that stage. Similarly, the present inquiry has not dealt with the welfare consequences of the agrarian ST, but mainly with its drivers and distributional consequences, among them its gender dimensions.

On an overall level, two of the three processes studied here do not seem to be gender biased. Most importantly, female farm managers are not excluded from grain intensification processes. Hopefully this implies that insofar as intensification has positive effects on farmer food security (directly through self-provisioning, or indirectly through increased farm incomes), women as well as men and their children would benefit from the process.[16]

[16] In a recent study, Ritzema et al. (2016) found moderate but positive effects of production intensification on smallholder food security.

Unlike the other two processes, in the most recent period commercial crop diversification is not gender neutral. Our model shows a clear discrimination in favour of MMFs. This gender bias could be connected to biases in some of the other drivers, since diversification is more likely to benefit farmers with access to land reserves, security of tenure, irrigation, and credit,[17] all factors with a positive influence on the odds for diversification.

Non-farm diversification does not seem to be a gendered process, at least at the household level: male- and female-headed households were as a likely to become pluriactive during the second period, although female-headed households had lower odds of non-farm diversification in the first period.

Since the elite is less non-farm diversified than the majority of smallholders, sector-straddling has become more of an option for the majority; it would be wrong to say that it is pro-poor but neither is it pro-rich. Interestingly, NFI to some extent seems to be invested in grain intensification, suggesting high returns in grain markets and possible food security improvements.

[17] Although binary statistics do not evidence discrimination of women in the access to agricultural credit.

6

Agricultural Intensification Response to Agricultural Input Subsidies in Tanzania: A Spatial-Temporal and Gender Perspective, 2002–15

Elibariki E. Msuya, Aida Cuthbert Isinika, and Fred Mawunyo Dzanku

Introduction

Attaining food security through self-sufficiency from local production has been a longstanding goal of the Tanzanian government (Isinika et al. 2005), and agricultural intensification has been the key strategy to achieve this goal in the main staple crops. Towards this end, many programmes implemented or coordinated by the government with support from numerous donors have come and gone, and have had varying levels of achievement as well as setbacks. The introduction of inorganic fertilizer and improved seed under the National Maize Project (1974–84) in the southern highlands of Tanzania has been cited as a successful example of agricultural intensification in Tanzania (Isinika et al. 2005) and even deemed a mini-green revolution by some (Eriksson 1993).

However, setbacks towards agricultural intensification have also been numerous. For instance, the mini-green revolution cited above could neither be replicated countrywide nor could it be sustained after market liberalization in 1986 due to high subsidy costs. All subsidies were subsequently withdrawn (World Bank 1994, Delgado and Minot 2000). Hence, low input use and correspondingly low agricultural productivity remains prevalent throughout the country. Consequently, poverty—particularly rural poverty—has persisted despite numerous anti-poverty development programmes and projects. The government has made various efforts to address these problems. Under the Agricultural Sector Development Programme (ASDP II), which contributes towards MKUKUTA,[1] targets had been set to increase the agricultural sector growth rate from 2.7 per cent in 2009 to 6.0 per cent per annum by 2015; reduce income poverty from 33.3 per cent in 2011/12 to 24.0 per cent by 2025; and and reduce food poverty from 11.3 per cent in 2011/12 to 5 per cent by 2015 (United Republic of Tanzania 2015).

This chapter assesses the trend of agricultural intensification in Tanzania, focusing on the performance of maize and rice[2] production from 2002 to 2015. Using panel data from the Afrint study (Tanzania), the chapter examines how maize and rice yields have changed in the Iringa and Morogoro regions during the study period, which coincides with the second phase of the agricultural subsidy scheme. It was envisaged by policymakers that providing an input subsidy would improve agricultural productivity and contribute towards the national goal for food security and poverty reduction (United Republic of Tanzania Ministry of Finance and Economic Affairs 2010). The main objective of this chapter is to assess whether the subsidy programme had an impact on reducing poverty as envisaged by the government, also comparing the yield outcomes and livelihood impacts across the gender, age, and wealth stratum. Specifically, the analysis addresses three questions: (1) What has been the effect of input subsidy programmes on agricultural technology adoption? (2) How has the adoption of technologies targeted by the subsidy programme impacted on maize and rice yields? (3) What has been the relationship between intensification (increased yield), output, and livelihood outcomes?

Data for the study were collected from two regions involving about 400 households in ten villages. All five villages from the Morogoro region belong to Kilombero district. The five villages in the Iringa regions are spread in three districts; Iringa rural, Kilolo, and Mufindi. Data were collected in three waves (2002, 2008, and 2015), as described in Chapter 1. In Afrint II and III, some of the original respondents who had passed away or had become too old were replaced by their married offspring, who had formed new households. By the final round of data collection, 186 of the original respondents remained. Out of these, 150 had complete data for all the relevant variables, thus constituting the panel data used in this study of temporal, spatial, and gender differences in use of inorganic fertilizer and improved maize and rice seed as productivity-enhancing technologies.

[1] MKUKUTA stands for Mpango wa Kukuza Uchumi na Kuondoa Umaskini Tanzania, in Kiswahili (the national language of Tanzania). MKUKUTA is the equivalent of the National Strategy for Growth and Poverty Reduction (United Republic of Tanzania, Ministry of Finance and Economic Affairs, 2010). It is an instrument for implementing the National Vision 2025 to guide economic transformation to attain middle income status by 2025.

[2] In this chapter rice is used interchangeably with paddy (un-milled rice).

Background

Agricultural Policy in Tanzania

Tanzania, like many other developing countries, aspires to attain economic development. Food security and poverty reduction are key indicators used to gauge success in meeting the national goal to attain mid-level economic development status by the year 2025, as stated in the National Vision (United Republic of Tanzania Ministry of Finance and Economic Affairs 2010). Indicators for achieving this goal are defined in the National Strategy for Growth and Poverty Reduction, more commonly known by its Kiswahili acronym—MKUKUTA. It has been argued by some authors that as long as the majority of the world's poor continue to derive their livelihoods from agriculture, any poverty reduction strategy must focus on improving agricultural performance because agricultural growth promotes poverty reduction (Ajani and Igbokwe 2013, Andersson Djurfeldt 2013). However, others have argued that agriculture is not a reliable sector on which to base rising prosperity since it has failed to deliver in the past. For instance, Ellis (2009: 59) argues that 'the notion that poverty reduction is best addressed in the very sector where poverty is most acute is seriously flawed and fails to focus adequately on supporting growth processes where they actually occur in the economy.'

In Tanzania, poverty has been defined as predominantly rural, and it is skewed against women and youths (United Republic of Tanzania 2015). More than 70 per cent of the rural population still depends on agriculture for their livelihoods, but increasingly also draws on non-farm income (NFI) sources (Katega and Lifuliro 2014, Ombakah 2014). At the macro level, the share of agriculture in GDP declined from 50 per cent in the year 2000 to less than 24.7 per cent in 2014 (United Republic of Tanzania 2008, Bank of Tanzania 2013/14), partly due to revision in the national account estimates during 2006 to accommodate structural changes in patterns of production, consumption, and investment. The revision also accounted for changes in relative prices and incorporated new data, which reflected new economic activities in the services, industry, and construction sectors (United Republic of Tanzania, National Bureau of Statistics 2006, Isinika et al. 2011).

At the farm level, such diversification is reflected by the rising share of non-farm economic activities, hence NFI (Da Corta and Price 2009). Nonetheless, farming remains the livelihood anchor for the majority of rural households in Tanzania for the foreseeable future. The Afrint III data show that nearly two-thirds of household income came from farming. Since the mid-1970s the government has used subsidies as the main instrument for promoting agricultural intensification in order to achieve food security and poverty reduction goals (Isinika and Msuya 2016). We therefore examine whether these targets have been met by recent subsidy programmes, and if they have been gender inclusive as intended.

Agricultural Subsidy Programme Since 2000

In Tanzania, we identify three phases of agricultural inputs subsidy provision. During the first round (1975–84) the inputs subsidy involved providing fertilizer, improved seed, and pesticides to farmers in Iringa, Mbeya, Rukwa, and Ruvuma, nicknamed 'the big four' (Isinika et al. 2005). This subsidy was phased out in 1984 due to high cost and realignment of policies to conform to a market economy (Delgado and Minot 2000). By 1994, all subsidies (direct and indirect) had been withdrawn. The second phase involved a partial transport subsidy that was introduced in 2003, lasting until 2007. This subsidy was introduced to address the problem of declining yield and output of maize in particular (Delgado and Minot 2000, Isinika et al. 2005). The subsidy was improved to accommodate seed for other crops (maize and sorghum in 2004) as well as other inputs (pesticide and herbicides in 2006), which were added to improve the subsidy's impact (Isinika and Msuya 2016). During the financial year 2007/8 the input subsidy programme was redesigned to form the National Agricultural Input Voucher Scheme (NAIVS), with improved administrative efficiency, constituting the third phase (2008–13). This study focuses on the second- and third-phase subsidy programmes.

The NAIVS scheme was developed as a pro-poor scheme to enhance smallholder contribution to increased production of maize and rice (United Republic of Tanzania 2015). The scheme intended to reach 2.5 million farmers by 2011, providing a 50 per cent subsidy for the purchase of chemical fertilizer as well as improved seed to increase food crop productivity and output, particularly for maize and rice in the big four high-potential regions, plus Kigoma and Morogoro, which were added (Patel 2011). In 2011 the subsidy coverage was expanded to cover sixteen regions, also extending it for another two years until 2013. According to the NAIVS project document, it was expected that by 2015 the scheme would have stimulated significant demand for inputs such that farmers' needs would continue to be met through the market. Nonetheless, by 2015, when data for Afrint III data were being collected, NAIVS had ended, but subsidized inputs were supplied to villages as new administrative arrangements were being organized to minimize corruption tendencies and inefficiency.

Methodology

Conceptual Framework

It has been argued that successful agricultural intensification should improve the productivity of land and labour (Larsson 2001), which improves food security at the household level and economic development community-wide, up to the district, regional, and national levels. Productivity from intensification also induces agricultural commercialization as farmers increase their share of marketed surplus over time (Jayne et al. 2011, Andersson Djurfeldt 2013). This, in turn, enables farmers to accumulate farm and non-farm assets, which contribute further towards improving farm productivity, while accumulated non-farm assets (houses, radios, and transport facilities) move the farmers to a higher wealth status. According to Ellis and Mdoe (2003), poverty-reducing accumulation involves trading up assets, for instance, from having a chicken to acquiring goats and later buying cattle and land. Another pathway may involve moving from owning chickens, which leads to having cash income, to financing non-farm activities. Similar indicators for assessing livelihood improvement are used in this chapter.

The pathway or trajectory from intensification to improved livelihoods, however, is not guaranteed, and it may vary from one farmer to another, based on choices farmers make or limitations they face in accessing such technologies. This means there will be farmers who use improved seed and inorganic fertilizer from subsidy sources and from the market. The group of non-users include those who did not access subsidy inputs because they could not afford to, and others who may have opted not to use the inputs due to other reasons. Close to one-third of the respondents (28 per cent) in the Afrint study could not afford to buy fertilizer (Msuya and Isinika 2016). The working hypothesis in this study is that users of improved inputs will have higher yields and better livelihood indicators compared to non-users.

Analytical Approach

The Afrint survey instrument was used to collect data on fertilizer and improved seed adoption. During each cross-section (wave) farmers were asked to report whether they used these inputs in the current period and for the two prior periods.[3] Mean values were then computed for the use of fertilizer and maize and rice seed. Data were also collected for maize and rice yield, representing the outcome of technology adoption. Information on asset ownership was used to compute an asset index to represent a household's welfare. These data are used for descriptive and regression analysis.

The analysis is done in two stages. Descriptive data are presented in Tables 6.1 to 6.5, comparing the mean percentage of farmers who chose

[3] During each cross-section, yield data were collected for the current year and two previous seasons. For example, a respondent who was interviewed in 2002 reported on maize yield for 2002, 2001, and 2000. The average of the three reported values is used throughout the analysis in this chapter.

Table 6.1. Mean difference in maize yields and agricultural technology adoption by region

Indicator	Morogoro	Iringa	Diff.
Yield (Kg/ha)	1,172	1,262	90 kg/ha
Seed (% of households using improved seed)	7	28	22***
Fertilizer (% of households using inorganic fertilizer)	4	43	39***
No fertilizer (% of households using no fertilizer)	89	46	–43***
Seed only (% of households using seed only)	6	11	5**
Fertilizer only (% of households using fertilizer only)	4	26	21***
Seed and fertilize (% of households using improved seed and fertilizer)	0	17	17***
N	297	399	

Notes: a. Unless otherwise stated, the definitions used for indicators used in Table 6.1 are also used elsewhere in this chapter.
b. Diff. in column four stands for difference.
c. ** and *** denote statistical significance at the 5 per cent and 1 per cent levels respectively.

different options on technology adoption and the corresponding mean yield for each cross-section or wave (Afrint I, Afrint II, and Afrint III). Descriptive analysis is done using pooled data, or waves comparisons being made by gender, region, and wealth quartile for maize. The analysis for rice basically applies to Morogoro since only a few farmers grew rice in Iringa. Figure 6.1 presents the relationship between yield and the farmers' welfare status represented by a wealth index, which was constructed using principle components analysis (Finan et al. 2005) based on ownership of various assets, including: phone, stove, television sets, radio, tape recorders, roofing material, and type of wall for houses.

The second level of analysis employed a fixed effects regression model to determine factors that influence the variation in yield of maize and rice in the Iringa and Morogoro regions. The analytical model is derived in the section on regression analysis. Findings from both analyses provide the basis to inform policymakers about the yield outcomes and livelihood impacts of agricultural subsidy programmes since their introduction in 2003. These findings can inform and influence the design and realignment of current and future subsidy programmes.

Findings

Descriptive Analysis

TECHNOLOGY ADOPTION YIELD RESPONSE AND THE SPATIAL GAP

Table 6.1 presents pooled data for the mean difference of adopting inorganic fertilizer and improved maize seed and corresponding mean yield values for Iringa and Morogoro. In the Morogoro region, only 7 per cent and

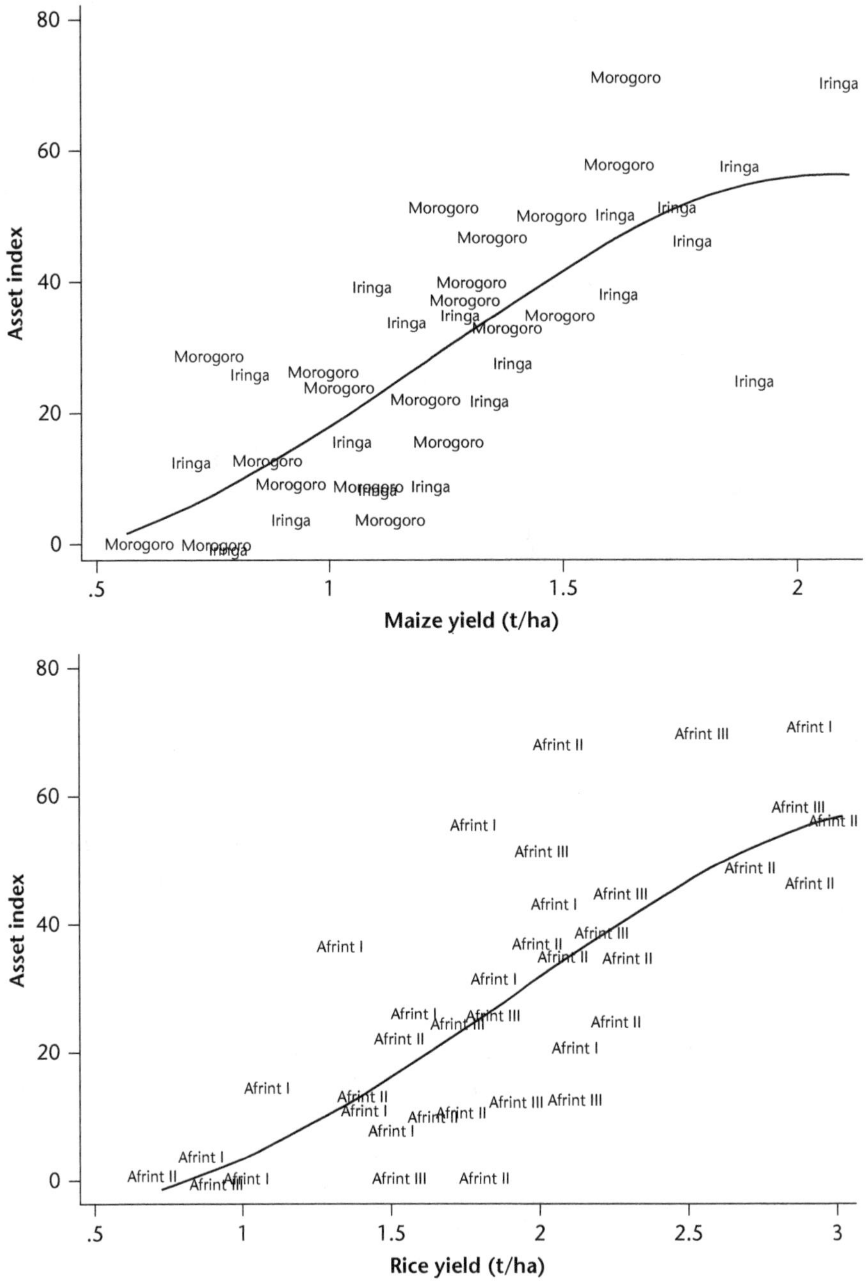

Figure 6.1 Welfare and agricultural intensification.

Notes: a. To make the graph more readable, each point represents thirty-five observations for the upper panel and twenty observations for the lower panel.

4 per cent of the respondents adopted improved seed and fertilizer respectively. Corresponding levels for Iringa were higher, being 28 per cent for seed and 43 per cent for fertilizer. The regional difference in adoption rates was much higher for fertilizer (39 per cent) compared to 22 per cent for seed; both differences were, however, significantly different from zero (hereafter simply stated as 'significant'). The level of technology adoption is much lower in Morogoro, where nearly 89 per cent of the respondents did not use fertilizer compared to 46 per cent in Iringa, a 43 per cent gap between the two regions.

The low level of fertilizer use in Morogoro may be explained by three factors. First, technologies for rice production in Tanzania are less developed and accessible relative to maize (Isinika et al. 2004). Second, low levels of improved seed adoption attract fewer farmers to adopt fertilizer application (Msuya and Isinika 2016). Third, rice production in the study area takes place in Kilombero valley, where soil fertility is relatively higher due to alluvial soils from annual flooding. During the third Afrint survey (2015) about 64 per cent of the respondents did not use inorganic fertilizer, the highest non-use being in Morogoro (83 per cent) where 53 per cent of the respondents felt their soil was fertile enough, compared to only 14 per cent in Iringa (Msuya and Isinika 2016).

Out of farmers who adopted any improved technology, in Morogoro 6 per cent adopted only seed, 4 per cent adopted only fertilizer, and very few farmers (approximately 0 per cent) adopted both seed and fertilizer. As stated earlier, farmers in Iringa had higher adoption rates: 11 per cent adopting only seed, 26 per cent only fertilizer, and 17 per cent adopting both seed and fertilizer. Mean yield levels were, as expected, higher in Iringa by 90 kg/ha, but the difference was not significant. This analysis shows clearly that for maize production there were spatial differences in technology adoption between the two regions, which may be attributed to higher soil fertility in Kilombero valley, but also due to the fact that the Morogoro region joined the subsidy programme later; rice seed was added in 2006 and fertilizer subsidy for rice production in Morogoro became available in 2008 (Patel 2011).

In Table 6.2 technology adoption and yield levels are compared for pooled data and for each region across all three waves. The difference between time periods is also presented. According to pooled data for maize, between Afrint I and Afrint II there was no significant difference in technology adoption but yield increased by 277 kg/ha, which was significant. The proportion of households adopting improved seed and fertilizer increased slightly, by 6 per cent and 3 per cent respectively. A slight increase was also observed for farmers who adopted seed only (1 per cent) as well as seed and fertilizer (4 per cent). But the proportion of farmers who did not use fertilizer declined by 4 per cent.

During the next interval (Afrint II to Afrint III), the proportion of farmers who adopted improved seed declined significantly by 8 per cent ($p \leq 0.05$) and

Table 6.2. Mean difference in yields and agricultural technology adoption over time

Indicator	Maize					Rice[†]				
	Afrint I	Afrint II	Afrint III	Afrint I–II	Afrint II–III	Afrint I	Afrint II	Afrint III	Afrint I-II	Afrint II–III
All regions										
Yield	1,002	1,279	1,390	277***	111	1,512	2,039	2,164	527***	125
Seed	17%	23%	16%	6%	−8%**	19%	21%	14%	2%	−7%
Fertilizer	20%	23%	35%	3%	12%***	12%	33%	28%	22%***	−5%
No fertilizer	69%	65%	60%	−4%	−5%	76%	54%	63%	−23%***	9%
Seed only	11%	12%	4%	1%	−8%***	12%	13%	9%	1%	−4%
Fertilizer only	14%	12%	24%	−2%	12%***	5%	25%	24%	21%***	−1%
Seed and fertilizer	7%	11%	12%	4%	0%	7%	8%	5%	1%	−3%
N	232	232	232			102	102	102		
Morogoro										
Yield	1,020	1,260	1,236	240**	−24					
Seed	4%	10%	6%	6%*	−4%					
Fertilizer	1%	1%	11%	0%	10%***					
N	100	100	100							
Iringa										
Yield	989	1,293	1,504	304***	212*					
Seed	28%	34%	23%	6%	−11%*					
Fertilizer	35%	40%	54%	5%	14%**					
No fertilizer	49%	46%	43%	−3%	−3%					
Seed only	16%	14%	3%	−2%	−11%***					
Fertilizer only	23%	20%	34%	−3%	14%**					
Seed and fertilizer	12%	20%	20%	8%*	1%					
N	133	133	133							

Notes: a. *, **, and *** denote statistical significance at the 10 per cent, 5 per cent, and 1 per cent levels respectively.
b. This applies to the Morogoro region only.

so did the percentage which adopted seed only (8 per cent; $p \leq 0.01$), both changes being significant. However, there was a significant increase in the proportion of farmers adopting fertilizer (12 per cent), but no change in those who adopted fertilizer and seed. Maize yield increased by 111 kg/ha (from 1,279 kg/ha to 1,390 kg/ha), but the change was not significant. During this interval (Afrint II to Afrint III) the proportion of fertilizer users in Morogoro region increased significantly by 10 per cent (from 1 per cent to 11 per cent), but the use of improved maize seed declined by 4 per cent contributing to a yield decline by 24 kg/ha, but the decline was not significant.

A similar pattern of changes is observed in the Iringa region in terms of technology adoption, but in this case significant improvements in yield are observed both for Afrint II and Afrint III, being stronger in the former period. In the Morogoro region, significant improvement in yield is observed during the first interval (Afrint I to Afrint II), attributed to increase in improved seed adoption, where 6 per cent more households used the technology but no significant change was observed in relation to fertilizer use. In the case of rice, significant improvement is observed in yield during the first interval when yield increased by 527 kg/ha, equivalent to 35 per cent. This is largely accounted for by the increase in households that used fertilizer (22 per cent). Households that used seed only or fertilizer and seed increased by only 1 per cent. During the second interval (Afrint II to Afrint III) the adoption of improved seed and fertilizer declined by 7 per cent and 9 per cent respectively, and the share of households that did not use fertilizer rose by 9 per cent. Nonetheless, paddy yield increased by 125 kg/ha.

The trend of technology adoption shows improvement during the first interval but a decline during the second interval; this can be explained by improvement of the subsidy programme whereby rice seed and the Morogoro region were added by 2008. The decline for Afrint III may be explained by the fact that the NAIVS scheme ended in 2014. During 2015, when the third wave of Afrint data was collected, the government was still deliberating about the future of agricultural inputs subsidy provision in terms of magnitude and institutional framework. Some subsidized inputs were distributed, from stocks of the expired NAIVS.

The NAIVS was designed to stimulate demand for improved seed and fertilizer, after which farmers would continue to get supplies from the market (Patel 2011). However, the programme evaluation report (United Republic of Tanzania 2015) shows that only 37 per cent of the farmers continued to use fertilizer beyond the three-year period. But, 50 per cent continued to use improved seed, indicating a higher preference to adopt improved seed than fertilizer. This last conclusion contradicts findings from the analysis in this study which show a higher proportion of farmers adopted fertilizer than improved seed.

THE GENDER GAP

Eligibility for the voucher programme was to be based on two hard criteria: (1) The household was cultivating no more than 1 ha (approximately 2.5 acres) of maize or paddy; and (2) the household could afford the top-up 50 per cent payment for the input package (approximately TZS[4] 55,000). In cases where the number of eligible households was larger than the number of vouchers, priority was to be given to female-headed households and farmers who had not been using improved inputs five years prior to NAIVS (Patel 2011). Table 6.3 presents the gender gap in technology adoption and yields for maize and rice. In the case of maize, the pooled data (covering all three waves) show that significantly more male-managed farms (MMFs) (10 per cent) used improved seed, but the proportion of female-managed farms (FMFs) using fertilizer was higher (by 5 per cent), but the difference was not significant. When the data are disaggregated by cross-section (wave) the dominance of MMFs using fertilize rises to 16 per cent during the first phase, but declines to 9 per cent and 5 per cent during the second and third phases respectively. The difference during the first phase is significant.

Despite this, MMFs consistently have significantly higher yields, as demonstrated by the significant difference between them for the pooled data as well as for each cross-section. The gender yield gap (269 kg/ha) is highly significant when pooled data are used. The mean yield of FMFs is only 79 per cent[5] of the mean for MMFs, thus a 31 per cent[6] gender gap. However, the mean yield of FMFs has caught up—as shown in the cross-sectional data. On average female farm managers attained 77 per cent of the yield of their male counterparts in Afrint II—a share that rose to 82 per cent for Afrint III. Between Afrint I and Afrint II, the yield of both MMFs and FMFs declined by 18 per cent and 20 per cent respectively. During this period the transport subsidy did not specifically target female farmers (*de facto, de jure,* or those within MMFs). The yield decline has been attributed to drought during the 2007/8 season. The rebound after 2008 was stronger for women (20 per cent) compared to men (7 per cent), which probably reflects the effect of NAIVS to specifically target FMFs (Patel 2011). Pooled data for rice show that significantly more MMFs applied fertilizer, a 13 per cent difference. However, the difference was not significant for improved seed (5 per cent) and yield (236 kg/ha); the mean yield of FMFs was equal to 88 per cent that of MMFs, clearly demonstrating the gender gap.

[4] TZS = Tanzania Shillings—the local currency for Tanzania.

[5] Computed using the mean yield of FMFs as a percentage of MMFs.

[6] Computed as the difference between the mean yield of MMFs minus the mean of FMFs over the mean of MMFs times 100.

Table 6.3. Mean difference in technology adoption and yields by sex of farm manager

	Maize				Rice			
	Yield	Seed	Fertilizer	*N*	Yield	Seed	Fertilizer	*N*
Pooled								
Male	1,281	21%	25%	548	1,945	19%	27%	254
Female	1,013	11%	30%	150	1,710	13%	13%	52
Diff.	269***	10%**	−4%		236	5%	13%**	
Afrint I								
Male	1,048	21%	20%	186	n.a.	n.a.	n.a.	
Female	815	4%	22%	46	n.a.	n.a.	n.a.	
Diff.	233**	16%***	−2%		n.a.	n.a.	n.a.	
Afrint II								
Male	1,354	25%	22%	184	n.a.	n.a.	n.a.	
Female	992	16%	27%	48	n.a.	n.a.	n.a.	
Diff.	362**	9%	−4%		n.a.	n.a.	n.a.	
Afrint III								
Male	1,452	17%	34%	176	n.a.	n.a.	n.a.	
Female	1,193	12%	39%	56	n.a.	n.a.	n.a.	
Diff.	260*	5%	−4%		n.a.	n.a.	n.a.	

Notes: a. *, **, and *** denote statistical significance at the 10 per cent, 5 per cent, and 1 per cent levels respectively.
b. n.a. = not available; values for these cells were less than 1 per cent, hence they are not displayed.

THE WEALTH RANK GAP

The national input subsidy scheme was designed to stimulate production, targeting poor farmers who cultivate 1 ha or less and women (Patel 2011). We argue that successful targeting of poor farmers should, over time, narrow the gap between the poor and the rich when it comes to the adoption of improved technologies targeted by the policy, all else being equal. We test this hypothesis using descriptive analytical tools where we compare fertilizer and improved seed adoption rates across wealth quartiles. We use an asset index as our measure of wealth.

Based on the $\chi 2$ test for the pooled data, results in Table 6.4 show that there was a significant difference between the four wealth ranks and so was the gap between the richest and the poorest quartiles; being highest for fertilizer use on rice (37 per cent) and lowest for use of improved rice seed (10 per cent). The technology adoption gap for maize was also wider for fertilizer used (17 per cent) than for seed (13 per cent). This could be explained by the fact that even though the price of improved seed is relatively higher compared to local varieties, seed normally requires smaller quantities per unit area, such that it is possible for farmers to buy the small quantities they can afford and plant on a small portion of their farm. Meanwhile, farmers have to apply relatively larger quantities of fertilizer per unit area for the impact to be felt.

Table 6.4. Agricultural technology adoption and yields by wealth quartiles

	Maize			Rice		
	Seed	Fertilizer	Yield	Seed	Fertilizer	Yield
Pooled						
Poorest	15%	20%	904	14%	5%	1,280
2nd	14%	22%	1,114	16%	18%	1,864
3rd	18%	27%	1,260	16%	34%	2,136
Richest	28%	36%	1,631	24%	42%	2,368
Overall statistical significance: $\chi^2(3)$	12.42***	12.03***	19.71***	3.65	22.02***	15.79***
Gap between poorest and richest	13%***	17%***	727***	10%	37%***	1,087***
Afrint I						
Poorest	9%	13%	830	7%	3%	1,122
2nd	24%	22%	1,088	14%	5%	1,830
3rd	18%	27%	1,147	26%	16%	1,750
Richest	21%	21%	1,354	30%	25%	2,078
Overall statistical significance: $\chi^2(3)$	5.57	3.70	5.07***	5.58	6.29*	9.21***
Gap between richest and poorest	12%*	8%	524***	23%**	22%**	956***
Afrint II						
Poorest	18%	25%	1,068	21%	7%	1,562
2nd	14%	12%	1,090	18%	36%	1,875
3rd	29%	22%	1,339	15%	41%	2,338
Richest	34%	34%	1,723	29%	54%	2,562
Overall statistical significance: $\chi^2(3)$	8.38**	7.76*	4.71***	1.65	13.46***	2.78**
Gap between richest and poorest	16%*	9%	655***	8%	47%***	999***
Afrint III						
Poorest	6%	29%	703	10%	7%	1,055
2nd	5%	30%	1,175	7%	30%	1,886
3rd	23%	38%	1,265	17%	28%	2,223
Richest	31%	47%	1,640	24%	52%	2,360
Overall statistical significance: $\chi^2(3)$	17.98***	4.52	17.10***	3.67	12.47***	11.81***
Gap between richest and poorest	25%***	18%*	937***	14%	45%***	1,305***

Note: a. *, **, and *** denotes statistical significance at the 10 per cent, 5 per cent, and 1 per cent levels respectively.

The difference in technology for improved maize seed was significant through all three waves of data collection but with increasing intensity, being 12 per cent in 2002 ($p \leq 0.1$), 16 per cent in 2008 ($p \leq 0.05$), and 25 per cent in 2015 ($p \leq 0.01$). The gap between all wealth ranks and between the poorest and richest households for fertilizer is significant for the pooled data but weaker for each cross-section. The proportion of households that used fertilizer increased only slightly (from 8 per cent to 9 per cent) between 2002 and 2008, but the overall difference between the wealth ranks was significant ($p \leq 10$ per cent) in 2008, while the gap between the poorest and richest stratum was significant in 2015. In the case of rice, the difference in adopting improved seed was strong only for the first wave (23 per cent). But the difference in adoption was stronger for using inorganic fertilizer on rice, which was significantly different for the pooled data ($p \leq 5$ per cent), and the intensity increased over time, being 22 per cent in 2003, 47 per cent in 2007 ($p \leq 0.05$), and 45 per cent in 2015 ($p \leq 0.01$). These findings imply that differences in technology adoption for maize are more strongly attributed to maize seed technology, while in the case of rice the push comes from fertilizer adoption.

Consistent with the technology adoption gap, there is a significant difference in the yield levels across wealth ranks both for pooled data (maize and rice) and for each wave of data collection ($\chi 2 > 2.78 - 19.71; p \leq 0.05 - 0.01$). The yield gap for pooled maize was 727 kg/ha, the poorest quartile obtaining only 55 per cent of the yield realized by the richest quartile. The corresponding ratio for rice is 54 per cent. The gap for maize increased from 524 kg/ha for the first round of data collection to 937 kg/ha by the third survey, rising by 413 kg/ha (79 per cent). Similarly, the gap for rice increased from 956 in Afrint I to 1,305 kg/ha by the time of Afrint III, which means the gap between the mean yield of the richest and the poorest quartiles increased by 349 kg/ha equivalent to 59 per cent. The widening gap, such that the richer quartile obtain higher yields implies that the richer quartiles were in a better position to utilize inputs supplied by the voucher system. Often, those now cultivate less than 1 ha, and farmers who had not engaged using inputs prior to 2003 cannot afford to pay 50 per cent of the price as required by NAIVS. For these reasons, criteria for accessing subsidized inputs had to be revised in some regions because very few farmers exceeded the proposed lower limit of 1 ha (Patel 2011).

AGRICULTURAL INTENSIFICATION AND LIVELIHOOD IMPROVEMENT

Everything else being equal, we expect increased agricultural intensification to lead to livelihood improvements. We examine this hypothesis by analysing the association between maize and rice yields and livelihood improvements using the asset index and a binary food insecurity status indicator (Table 6.5). A household is defined as food insecure if the number of meals eaten during the lean season is less than that eaten during the rest of the year. The food insecurity variable is available in this dataset for the second and third waves only. We begin by showing the correlation between intensification and welfare in Figure 6.1. The analysis clearly shows that the welfare of a household increases with rising yield level, but at a declining rate above 2 t/ha for maize and 3 t/ha for rice. Both yield levels fall short of recommended optimum levels

Table 6.5. Association between maize and rice intensification and livelihood outcomes

	Maize		Rice	
	Average welfare	Proportion food insecure	Average welfare	Proportion food insecure
Pooled				
Lowest	20.0	36%	17.9	26%
2nd	24.6	22%	27.8	34%
3rd	30.7	26%	35.3	27%
Highest	37.2	24%	39.5	33%
Overall significance: *F*-stat(3) and $\chi^2(3)^b$	24.03***	5.75	15.85***	1.24
Gap between highest lowest	17.2***	−12%	21.6***	7%
Afrint I				
Lowest	11.3	n.a.	6.3	n.a.
2nd	16.7	n.a.	16.8	n.a.
3rd	18.5	n.a.	30.5	n.a.
Highest	24.6	n.a.	28.5	n.a.
Overall significance: *F*-stat(3) and $\chi^2(3)^b$	8.09***	n.a.	22.84***	
Gap between highest lowest	13.4***	n.a.	22.2***	n.a.
Afrint II				
Lowest	22.2	14%	21.2	0%
2nd	23.9	9%	23.5	12%
3rd	28.4	9%	28.2	15%
Highest	32.8	7%	33.9	24%
Test for overall significance: *F*-stat(3) and $\chi^2(3)^b$	4.95***	1.61	2.30*	1.30
Gap between highest lowest	10.6***	−7%	12.6**	24%**
Afrint III				
Lowest	31.1	59%	33.1	73%
2nd	35.9	35%	45.0	36%
3rd	44.2	44%	45.0	42%
Highest	47.7	37%	49.4	40%
Overall significance: *F*-stat(3) and $\chi^2(3)^b$	11.51***	8.51**	2.96***	8.71**
Gap between highest and lowest	16.6***	−22%**	16.2***	−33%**

Notes: a.*, **, and *** denote statistical significance at the 10 per cent, 5 per cent, and 1 per cent levels respectively.
b. The F-stat (3) is for the test related to the asset index while the $\chi^2(3)$ is of the binary food insecurity indicator.
c. n.a. = values are too small (less than 1 per cent).

(4 t/ha for maize and 6–8 t/ha for rice) (Isinika et al. 2004), which means there is still potential to improve yield further.

Table 6.5 compares the proportion of farmers in a particular wealth rank and the corresponding share of food insecure households. For pooled data among farmers who grew maize, the lower ranks accommodate a higher proportion of food insecure households. While two of the lowest quartiles account for 44.6 per cent of the households producing maize, they constitute 58 per cent of the food insecure. Similar proportions apply for rice, where 45.7 per cent of the lowest quartiles account for 60 per cent of the food insecure.

The overall variation in welfare as well as food insecurity among wealth ranks is significant. This difference persisted through all three waves but the difference in food security is only significant during the third wave. The trend for rice production follows a similar pattern.

Regression Analysis

MODEL

The regression equations to be estimated are concerned mainly with assessing the difference in maize and rice yields between households using improved farm technology and those that are not. Because our interest is in government-subsidized inputs we focus mainly on inorganic fertilizers and improved seed adoption as the main explanatory variables, while controlling for other standard factors that typically enter an agricultural production function. Let Y_{kjit} represent output per unit area for crop *k* cultivated in region *j* by household *i* in time *t*, Z_{kjit} is the corresponding vector of production inputs (including inorganic fertilizers and improved seeds). Suppose X_{kjit} is the vector of all other covariates observed by the researcher that are assumed to influence crop output. Given the above assumption we can specific the yield function as:

$$Y_{kjit} = \delta' Z_{kjit} + \beta' X_{kjit} + u_i + e_{it}, k, j = 1, 2; i = 1, 2, ..., N; \quad \text{and} \quad t = 1, 2, 3 \tag{6.1}$$

where u_i represents an unobserved household-specific factor assumed to be time invariant, and e_{it} is the random error that captures exogenous shocks like weather variability.

Since the level of output obtained depends on farm household *i*'s choice of whether or not to use production inputs, and what level to use (based on observed and unobserved characteristics of the household and farm), Z is correlated with u_i *ex ante*. This means that to estimate δ' consistently we need an estimator that allows such correlation. The fixed effects (FE) estimator is thus the preferred estimator for evaluating the parameter vectors (δ' and β') in equation (6.1) because it allows both Z and X to be correlated with u_i. Arguably, the downside is that we are unable to identify time-invariant covariates that may influence yields (e.g. sex of farm manager). In the present application, however, the sex of farm manager contains appreciable variability over the approximately thirteen-year span of the panel (2002–15).

In order to allow us to test the differential impacts of technology use as a package, and given that our main variables of interest are captured in the binary form, we construct a discrete choice variable, *S*, which takes on the value 1 if a household neither used improved seeds nor applied inorganic fertilizers, 2 if a household used improved seeds but not inorganic fertilizers, 3 if a household used inorganic fertilizers but not improved seeds, and 4 if a household used both improved seeds and inorganic fertilizers.[7] Our hypothesis is that the yields obtained by farmers using a combination of the two inputs is over and above the yields of both those using only one of these inputs and those using none at all. Also, households using at least one of these inputs are expected to have higher average yields than those not using any of the two inputs at all. With this modification we can rewrite equation (6.1) as:

$$Y_{kjit} = \delta_2 S_{2kjit} + \delta_3 S_{3kjit} + \delta_4 S_{4kjit} + \gamma' Z_{kjit} + \beta' X_{kjit} + u_i + e_{it} \tag{6.2}$$

where δ_2, δ_3, and δ_4 are the coefficients on the discrete choice input variable of interest described above, meaning that δ_1, the coefficient on S_1 (i.e. not using any of the two inputs) is the reference category. Based on our hypothesis, we expect δ_2, δ_3, and δ_4 to be positive and statistically different from zero. The vector γ' represents the coefficients of other inputs such as cultivated area, labour, machinery, and other soil amendments.

In the case of maize, equation (6.2) is estimated on the full sample and also separately for the two regions (i.e. Morogoro and Iringa). Rice is predominantly produced in Morogoro and therefore the rice equation estimates apply to that region only.

RESULTS OF REGRESSION ANALYSIS

Results of regressing the mean yield against the mean values of variables discussed in the previous section are presented in Tables 6.6 and 6.7 for maize and rice respectively. In the case of maize, Table 6.6 presents results of analysing the full sample data as well as the two regions separately. Each of the models was tested for overidentification using the Sagan Hansen statistics, and all $\chi 2$ values were found to be highly significantly greater than zero, implying that the fixed-effect model was suitable for statistical inference. All three models had F values that were highly significant for each of the three waves. The adjusted R^2 increased from 14.4 per cent to 18 per cent and 20.6 per cent during the three waves of data collection, implying that the combined effect of explanatory variables to account for variation in the mean maize yield increased over time.

The intercept was positive and highly significant for all three waves. Most of the coefficients are positive, except for the natural log of cultivated area, which is negative and highly significant ($p \leq 0.01$), confirming the longstanding inverse relationship between farm size and productivity (Muyanga and Jayne 2014). This is consistent with the argument that in Tanzania, after the year 2000, crop production response has more often come from area expansion rather than productivity improvement (Delgado and Minot 2000).

[7] Here, we suppress the subscripts k, j, i, t.

Table 6.6. Fixed-effect estimates of the determinants of maize yields with improved seed and fertilizer adoption effect

Variables	Overall			Morogoro region			Iringa region		
	(1)	(2)	(3)	(1)	(2)	(3)	(1)	(2)	(3)
Improved seed and fertilizer (ref. = neither)									
Seed only	−0.048	−0.044	−0.023				−0.073	−0.064	−0.013
	0.093	0.093	0.095				0.115	0.118	0.123
Fertilizer only	0.178*	0.144	0.112				0.168	0.126	0.080
	0.097	0.089	0.090				0.111	0.097	0.099
Seed and fertilizer	0.260*	0.210	0.163				0.252*	0.198	0.160
	0.133	0.128	0.127				0.140	0.133	0.129
Improved seed				0.025	0.034	−0.005			
				0.153	0.147	0.140			
Fertilizer				0.229	0.185	0.173			
				0.205	0.211	0.199			
Log area cultivated	−0.332***	−0.359***	−0.333***	−0.434***	−0.416***	−0.406***	−0.282***	−0.325***	−0.283***
	0.076	0.079	0.075	0.147	0.151	0.135	0.085	0.089	0.089
Log labour use	0.098	0.072	0.059	−0.018	−0.009	−0.021	0.160**	0.123*	0.118*
	0.060	0.059	0.059	0.121	0.120	0.115	0.068	0.067	0.067
Soil amendments	0.059***	0.059***	0.060***	0.065*	0.067*	0.072*	0.058**	0.057**	0.057**
	0.022	0.022	0.022	0.037	0.036	0.036	0.028	0.027	0.028
Agricultural extension	0.107*	0.081	0.073	0.135	0.109	0.055	0.098	0.073	0.079
	0.060	0.060	0.059	0.088	0.083	0.080	0.083	0.084	0.081
Mechanization	0.114**	0.091	0.039	0.103	0.085	0.085	0.107	0.084	0.046
	0.056	0.058	0.060	0.073	0.079	0.086	0.086	0.090	0.089
Female manager		−0.119	−0.156		0.103	0.097		−0.160	−0.206**
		0.096	0.095		0.186	0.194		0.105	0.100
Age of manager		0.004	−0.001		0.000	−0.002		0.006*	0.001
		0.002	0.003		0.004	0.004		0.003	0.004
Educated manager		0.326***	0.220**		0.261*	0.199		0.346***	0.213
		0.094	0.098		0.140	0.138		0.120	0.129
Non-farm participation		−0.019	−0.015		−0.019	−0.001		−0.032	−0.037
		0.053	0.052		0.094	0.087		0.065	0.065

Time effect (ref. = 2002)									
2008			0.146***			0.264***			0.052
			0.050			0.070			0.070
2015			0.212***			0.176*			0.216***
			0.055			0.090			0.069
Constant	6.490***	6.210***	6.429***	6.488***	6.331***	6.384***	6.526***	6.170***	6.422***
	0.068	0.158	0.171	0.121	0.242	0.233	0.089	0.204	0.225
Observations	696	695	695	297	296	296	399	399	399
Adjusted R^2	0.144	0.180	0.206	0.138	0.163	0.220	0.155	0.206	0.229
F-stat.	9.005	7.555	7.971	4.704	3.310	4.522	5.400	4.833	6.057
p-value of *F*-test	0.000	0.000	0.000	0.000	0.000	0.000	0.000	0.000	0.000
sigma_u	0.569	0.554	0.555	0.600	0.587	0.585	0.536	0.513	0.524
sigma_e	0.516	0.508	0.501	0.498	0.497	0.483	0.534	0.522	0.516
rho	0.549	0.543	0.551	0.591	0.582	0.595	0.502	0.492	0.507

Test of overidentifying restrictions: fixed vs random effects with cluster robust standard errors

	(1)	(2)	(3)	(1)	(2)	(3)	(1)	(2)	(3)
Sargan-Hansen statistic	50.391 $\chi(8)$	48.005 $\chi(12)$	49.293 $\chi(13)$	24.842 $\chi(7)$	27.872 $\chi(11)$	29.894 $\chi(12)$	59.073 $\chi(8)$	55.591 $\chi(12)$	61.277 $\chi(14)$
[*p*-values]	[0.0000]	[0.0000]	[0.0000]	[0.0008]	[0.0034]	[0.0029]	[0.0000]	[0.0000]	[0.0000]

Notes: a. Cluster robust standard errors are reported under coefficients.
b. ***, **, and * denote statistical significance at the 1 per cent, 5 per cent, and 10 per cent levels respectively.
c. sigma_u is the estimate of the within-household standard deviation.
d. sigma_e is the estimate of the between-household standard deviation.
e. rho is the estimate of the intra-class correlation or the proportion of the total variance due to heterogeneity across households.

Table 6.7. Fixed-effect estimates of the determinants of rice yields in Morogoro with improved seed and fertilizer adoption effect

Variables	(1)	(2)	(3)
Improved seed and fertilizer (ref. = neither)			
Seed only	0.016	−0.017	−0.053
	0.122	0.130	0.122
Fertilizer only	0.345**	0.306**	0.111
	0.137	0.138	0.143
Seed and fertilizer	0.459***	0.439***	0.326**
	0.137	0.140	0.134
Log area cultivated	−0.392***	−0.419***	−0.378***
	0.123	0.126	0.124
Log labour use	−0.089	−0.108	−0.100
	0.106	0.107	0.104
Agricultural extension	0.227**	0.200**	0.149*
	0.089	0.088	0.084
Mechanization	0.349***	0.331***	0.266***
	0.056	0.064	0.070
Female manager		−0.113	−0.129
		0.221	0.223
Age of manager		0.003	−0.001
		0.004	0.004
Educated manager		0.235*	0.166
		0.130	0.138
Non-farm participation		−0.073	−0.057
		0.082	0.082
Time effect (ref. = 2002)			
2008			0.238***
			0.081
2015			0.270***
			0.088
Constant	7.071***	6.856***	7.046***
	0.092	0.214	0.233
Observations	306	305	305
Number of households	102	102	102
Adjusted R^2	0.229	0.255	0.302
F-stat.	9.921	7.161	6.797
p-value of *F*-test	0.000	0.000	0.000
sigma_u	0.506	0.505	0.491
sigma_e	0.475	0.470	0.458
rho	0.532	0.536	0.536

Test of overidentifying restrictions: fixed vs random effects with cluster robust standard errors

	(1)	(2)	(3)
Sargan-Hansen statistic	34.712 $\chi(7)$	47.481 $\chi(11)$	42.780 $\chi(12)$
[*p*-values]	[0.0000]	[0.0000]	[0.0000]

Notes: a. Cluster robust standard errors are reported under coefficients.
b. ***, **, and * denote statistical significance at the 1 per cent, 5 per cent, and 10 per cent levels respectively.
c. sigma_u is the estimate of the within-household standard deviation.
d. sigma_e is the estimate of the between-household standard deviation.
e. rho is the estimate of the intra-class correlation or the proportion of the total variance due to heterogeneity across households.

Adopting only improved seed also has a negative sign through all three waves, indicating that for good performance improved maize seed requires fertilizer as a complementary input. This means that when farmers adopt technologies, they need to consider their complementary effects; otherwise, technology adoption may have a negative influence on yields. Adopting fertilizer only and combining seed and fertilizer both had a significant effect on yield, and while the influence declined during subsequent waves it remained positive. The findings also show that female managers recorded lower yields, consistent with the finding that FMFs use less yield-enhancing technologies, as reported earlier. The negative coefficient on non-farm participation reflects competition for resources, especially labour and capital.

Variables that had a significant positive effect on yield ($p \leq 0.1 - p \leq 0.01$) included: soil amendments, mechanization, labour use in the Iringa region, agricultural extension, education of farm manager, and time; these were stronger for the full sample and Iringa region than Morogoro. Soil amendments were similarly stronger in Iringa than Morogoro. The model for rice exhibits similar characteristics, having highly significant F values with an R^2 value that increased from 22.9 per cent in 2002 to 30.2 per cent in 2015. All the signs are similar to the maize model except for the negative sign on labour use, reflecting competition for this input among enterprises.

Conclusion and Policy Recommendations

The analysis presented in this chapter clearly shows that maize farmers in Iringa and Morogoro were highly responsive to the fertilizer and seed components of the input subsidy programme, and their response was sensitive to the magnitude of the subsidy. The partial transport subsidy, which was introduced in 2003, did not stimulate a strong yield response from farmers. After 2008, yields improved significantly, especially for rice but also for maize, which coincided with the introduction of the national subsidy scheme that was designed to stimulate production targeting poor farmers. In the absence of the subsidy, use of improved inputs for rice production in Morogoro had been low, especially among FMFs and households in the lowest production quartile. These categories should therefore be supported to improve productivity so that they can also make substantial productivity improvements and livelihood gains from government programmes. The wide gender gap in yield and livelihood indicators also needs to be addressed as stated in government policy documents.

7

Intensification, Crop Diversification, and Gender Relations in Malawi

Peter Mvula and Wapulumuka Mulwafu

Introduction

Agricultural intensification is currently being promoted in Malawi as one of the strategies for achieving food security while at the same time improving rural livelihoods. Continual cases of food shortages, vulnerability, and poverty lend credence to the narrative of 'low maize productivity trap' which has characterized Malawi's agriculture since the 1980s (Devereux 2002, Chirwa and Dorward 2013). In view of the declining levels of productivity and challenges of land shortages, agricultural intensification and crop diversification are some of the ways in which rural livelihoods can be improved. This chapter builds on the idea of intensification by analysing the dynamics of agricultural production in the rural setting of Malawi. It focuses on two key issues: the gendered patterns of intensification and the diversification of crop production. The study sought to answer the following research questions: What factors are driving agricultural intensification in the Afrint study areas in Malawi? How is agricultural intensification affecting both male and female farm managers in the Afrint study areas? What is the relationship between agricultural intensification and crop diversification in the study areas?

Data for the writing of this chapter were drawn from the three rounds of the study. The research was conducted in four regions, namely: Ntchisi, Thiwi Lifidzi, Shire Highlands, and Bwanje Valley. For this chapter, we used both the quantitative dataset and qualitative data. Qualitative data are only from the period of the third round of data collection and were only collected from three villages in two regions based on changes in the dataset between Afrint I and Afrint II. The three villages (here anonymized) are Chikwanje and Nkhwangwa in Ntchisi and Khasu in Thiwi Lifidzi. A detailed description of the approach, sample design, and data collection methods and analysis has been provided in Chapter 1.

In this chapter, we analyse differences between households with male farm managers and those with female farm managers. We use the gender of farm managers because they are often the ones who make decisions regarding what to grow, how to grow it, and what to sell. In our sample, the proportion of male farm managers did not differ very much from the proportion of households that were headed by men, and the same goes for female-headed households. Among female-headed households, 99.2 per cent had female farm managers and only 0.8 per cent had male farm managers. The opposite is true among male-headed households, where 97.5 per cent had male farm managers and only 2.5 per cent had female farm managers. This implies that more often than not, household heads were also farm managers.

The first part of the chapter provides an overview of the general agricultural policies pursued by the Malawi government since independence. Special attention is placed on demonstrating the aspects of intensification and diversification with regard to gender relations. The next section presents findings from the study and shows ways in which gender relations resonate with policy and practices when promoting intensification and crop diversification.

Background

The role of agriculture in Malawi's national development endeavours cannot be overemphasized. The sector contributes significantly to food security and rural livelihoods, employment, income, and exports. For instance, agriculture accounts for nearly 80 per cent of Malawi's export earnings, contributes a share of 36 per cent of its GDP, and occupies 80 per cent of the workforce (Mwase et al. 2014). Despite the centrality of this sector, efforts towards intensification and diversification have had no effects on rural livelihoods (Chirwa et al. 2008). Indeed, although Malawi has followed the spirit of the Maputo declaration to the letter by consistently allocating more than 10 per cent of the national budget to the agricultural sector, the problem of food shortages has not disappeared. The recent postcolonial history of Malawi has been characterized by frequent famines and food shortages. Many questions have been asked as to why the country continues to experience such disasters when the Ministry of Agriculture has traditionally received a significant amount of budgetary allocation compared to other sectors. Indeed, a lot of investment has been made in food production, raising incomes for farmers, soil fertility improvement programmes, and technology transfers. However, the impact has been adversely affected by persistently poor harvests and increasing impoverishment since the 1980s. In part, this is due to the country's overreliance on rain-fed agriculture as well as low technology uptake by smallholder farmers (Chirwa et al. 2008). This is manifested in the recurrent food shortages, the dwindling export base, and the flood of agricultural imports which have severely depressed local production (Harrigan 2003, Peters 2004b).

Chirwa and Dorward (2013) have suggested three phases in understanding the development of Malawi's postcolonial agricultural policies. The first phase, spanning the period from 1964 through 1980, is described as pre-reform and was marked by state regulation of the agricultural industry. The state intervened in areas of production, extension, technology development, and marketing of agricultural produce. Alongside donor support, government made significant investments through the integrated rural development programme. Input subsidies were also made while a state parastatal dominated in the marketing of peasant-produced crops. Both men and women participated in different agricultural activities and in some cases they were organized in farmers' groups in order to access inputs.

The second phase was ushered in by the adoption of Bretton Woods structural adjustment programme conditionalities in the early 1980s. The agricultural sector underwent some major reforms including the removal of subsidies and liberalization of trade. The real turning point occurred in the 1990s when, as a result of liberalization of trade and the withdrawal of subsidies, the agricultural sector virtually collapsed. The Agricultural Development and Marketing Corporation's (ADMARC) functions of cushioning and moderating input and producer prices to the advantage of smallholder farmers were withdrawn from the agricultural sector, and this served as a deterrent to production. In addition, the private sector began to play a role even though its performance has not been very satisfactory, and cases of abusing the smallholder farmers have been reported (Chirwa 1998). Moreover, the market information system was inadequate to address farmers' needs. Finally, the prices of farm inputs were too high for the smallholder farmers. By 1994, however, agricultural productivity remained relatively low as poverty and vulnerability were exposed by the food shortages of 1987 and 1992–4. The poor members of society, including women, suffered most under the effects of these reforms. But the reforms also revealed structural weaknesses in the economy, including the slow growth in exports, the lack of diversification, and overreliance on tobacco exports (Chirwa et al. 2008).

The third phase is the post-reform, covering the period after 1994 to the present. Agricultural policies developed during this period reflect the political climate of competitive politics where increasing agricultural productivity is necessary for both achieving food security and for enhancing political patronage. It is a phase sometimes marked by inconsistencies in policy formulation and implementation. One issue that has attracted a considerable amount of debate is that of the input subsidy programme. Before liberalization in the 1990s, the government used to run an input subsidy programme, particularly for fertilizer and seed. Although the programme was meant to stimulate smallholder maize production, it benefited the estate sector as well. For instance, large-scale commercial farmers could buy fertilizer at the same price as the smallholder farmers. The unfortunate part is that smallholder farmers were not allowed to grow high-income crops and remained with maize production. There was little diversification in terms of the crops grown. The greatest beneficiaries of the subsidy were therefore producers of high-income crops such as tobacco. However, in the 1994/5 growing season, Malawi liberalized the fertilizer market by introducing competition into the domestic smallholder market and reducing the monopoly of the Small Farmer Fertilisers' Revolving Fund of Malawi and ADMARC. The result was that fertilizer prices went up so high that not many farmers could afford to buy them. Consequently, production reduced drastically and Malawi began to experience food shortages in the mid-1990s, and the country began to import maize in order to meet the national food requirements (Chirwa and Dorward 2013).

In order to mitigate the effects of food shortages, the government intervened by introducing the Starter Pack Programme in the 1994/5 farming season which was later substituted by a Targeted Input Programme (TIP) in the 1996/7 farming season, whereby small quantities of packaged seed and fertilizer were distributed to needy farmers. Women, the elderly, and persons living with HIV/AIDS were particularly earmarked for this scheme. However, the subsidy was too little to have any meaningful impact on the country's food security. In addition, emphasis was largely on maize production. The food shortage situation was never completely addressed, as evidenced in the hunger crisis of 2002 (Devereux 2002, Dorward et al. 2004). The Agricultural Productivity Investment Programme (APIP) was another intervention in the 1990s aimed at increasing maize productivity. Funded by the European Union, this programme provided subsidized credit scheme facilities to smallholder farmers to increase maize productivity. However, by the early 2000s it was established that smallholder production of maize was inadequate to enable farmers to repay the loan on the basis of maize alone. Hence, diversification strategies were introduced to encourage the production of legumes (Government of Malawi 2007, Chirwa et al. 2008).

Literature on land tenure and the role of women in agricultural production and the differential access to resources is abundant (Kishindo 2004, Kerr 2005, Peters and Kambewa 2007, Peters 2010). As is the case in many parts of Africa, women perform most of the production tasks ranging from planting to harvesting of crops. At the same time, they attend to the various household chores like cooking, fetching water, and caring for children. While land in the study areas is traditionally owned by women, decisions on the use and management of produce is controlled by men. Women face further constraints to production including labour and technology, as both men and women depend on farming for their livelihoods but men tend to have greater opportunities of participating in wage employment and non-farm income activities.

Intensification Policies and Gender Dimensions

The state has remained an important actor in efforts to increase agricultural productivity and to achieve food security. For instance, in 2006, the government came up with the Agricultural Development Programme whose aim is to harmonize the investment and support programmes in agriculture which have the highest potential for contributing to food security and agricultural growth in the next five years. It is a prioritized, results-based framework for implementing agricultural components of the Malawi Growth and Development Strategy (MGDS). It has identified key constraints of the agricultural sector and the required investments within the context of national and regional contexts. It is essentially an operational tool of the MGDS in the area of agriculture, food security, irrigation, and disaster risk reduction. On agriculture, the idea is to increase the contribution of the sector to economic growth through production of food crops and value added for domestic and export markets (Government of Malawi 2007).

Intensification has been promoted as one of the strategies for addressing the challenge of low agricultural productivity. The introduction of input subsidy programmes, although in a modified form from previous initiatives, was aimed at increasing productivity of principally the maize crop. In the 2004/5 farming season, government introduced the farm input subsidy whereby farmers were provided with coupons to buy hybrid seed and fertilizer at subsidized prices. Although heavily criticized by some donors, the programme registered an increase in production to the point of temporarily making Malawi a self-sufficient nation. As in previous subsidy programmes, women and other vulnerable groups were the main targeted beneficiaries (Levy and Barahona 2002, Chinsinga and O'Brien 2008, Chirwa and Dorward 2013, Mwase et al. 2014).

Malawi's agricultural development has been driven by several factors including the promotion of irrigated agriculture, the subsidy programme, and the opening up of rural markets. One of the interesting aspects of liberalization has been the emergence of elite private groups trading in agricultural commodities. While the participation of the private sector has in some cases been instrumental in providing market opportunities in the rural and often hard-to-reach areas (Sitko and Jayne 2014a), the situation in Malawi shows that private traders have generally been exploitative. Civil servants and businessmen have taken advantage to establish small-scale companies, sometimes 'briefcase' companies for the purpose of participating in the market opportunities opened up with liberalization. The problem is that they have squeezed out the small farmers since they have knowledge of the markets (Chirwa et al. 2008).

Private traders have been quite active in the marketing of produce. Again, the most accessible areas have benefited from private sector participation. For input supply it is mainly fertilizer, and this again has been controversial. There have been complaints of delays and selectivity in the areas private traders target. Sometimes this has been attributed to political influence and not market principles of demand. Chirwa (1998) has argued that although the gains and losses are not clear-cut, the major losers of these policy reforms are smallholder farmers who buy maize at high and volatile prices. These net food buyers are generally low-income or wage earners in urban and semi-urban areas and smallholder farmers in remote areas. These groups have suffered as a result of increased consumer prices and seasonal price instability for major food crops. Chirwa (1998: 91) further argues that 'remote areas which cannot be accessed by private traders are losers because most of the markets in such areas were also closed by the state marketing agency. The closure of markets meant that net food buyers did not have access to surplus maize while net food sellers had difficulties in selling the surplus maize to private traders at better prices.'

More recently, government implemented the Irrigation, Rural Livelihoods and Agricultural Development (IRLAD) project which promoted some aspects of intensification. Funded by the International Development Agency, the US$65 million project sought to raise agricultural productivity of rural households by providing an integrated package of support covering irrigation, production advisory services, and marketing between 2006 and 2012. With more than 40 per cent of the beneficiary households being female-headed, IRLAD intensified production of maize and rice as a way of achieving food security. At the end of the implementation period, the project contributed towards increasing farm income and building of institutional capacity for irrigation development in several districts of the country (Government of Malawi 2012, Government of Malawi 2015). The results of this project run contrary to the findings of Ricker-Gilbert et.al (2014), who argue that while areas of higher population density are associated with smaller farm sizes, there is no evidence of intensification. They show that households have not experienced any increase in maize yields per hectare, but instead tend to rely more on off-farm income to earn a living.

Diversification Strategies and Gender Dimensions

Although the debate on agricultural diversification in Malawi has been going on for decades, the structure of the country's economy has remained relatively the same (Orr and Mwale 2001, Simtowe 2010, Chibwana et al. 2012, Asfaw et al. 2015). Concern for the country's overreliance on tobacco and maize has been expressed by a diverse group of stakeholders including government, donors, and partners. At the national level, initiatives have been introduced with qualified success. The European Union supported government with €23 million (approximately MK4.6 billion[1]) for the Farm Income Diversification Programme (FIDP) from 2009 to 2015 (Zant 2012). The recently ended phase one aimed at increasing food security and income levels of rural households, while at the same time ensuring sustainable use of soil and water resources by encouraging agribusiness development and improved marketing of agricultural products. It focused on the areas of soil conservation and water management, organization of rural communities, promotion of agribusiness initiatives, and improvement of capacities of rural communities to access and develop post-harvest agricultural activities and provision of timely and relevant training in horticulture. The second phase of the programme aims at improving the livelihoods and nutritional status of rural households through increased and diversified production, and better market access. Having learnt lessons from phase one, FIDP II uses an integrated approach and seeks to address the need: to diversify and increase agricultural production; to promote income generation through agri-business initiatives with the involvement of the private sector; to improve FIDP farmer group management capacities; to improve FIDP farmer group social dynamics; to halt and reverse the decline of the natural resource base; and to capacitate farmers and district level staff on nutrition issues (Zant 2012).

The declining natural resource base and climate variability have been viewed as some of the factors influencing farmers' adoption of crop diversification. The report by Asfaw et al. written for the FAO (2015) shows that climate variability is one of the main drivers of diversification in developing countries. In the context of Malawi, the study showed that climate variability determined the likelihood of diversification in terms of labour, cropland, and income. In terms of gender, women were less likely to diversify their labour than men. This means that as a risk-management and shock-coping strategy, diversification would afford male farmers more opportunities to increase income security than women farmers.

[1] MK stands for Malawi kwacha, the local currency.

At the national level, Malawi has several alternative crops to replace or supplement the traditional ones like maize, tobacco, sugar, and tea, but whose market outlook looks poor, partly due to high transportation costs. Yet the need to diversify production has become necessary in order to optimize returns to the land and labour. The major constraint to intensification and diversification in Malawi include high transportation costs and market infrastructure. In the 1990s, paprika was unsuccessfully promoted as one of the cash crops to diversify Malawi's economy. In more recent years, the food crops being earmarked for diversification include cassava, sweet potatoes, Irish potatoes, pigeon peas, and bananas, while cash crops include groundnuts, soya beans, macadamia, and paprika (Zant 2012, Asfaw et al. 2015, CISANET Nd).

Studies on diversification in Malawi reveal structural challenges for its large-scale uptake. In his study of the determinants of livelihood diversification in Malawi, Simtowe (2010) argued that female-headed households were more likely to combine agriculture with casual labour than merely relying on agriculture. In addition, access to credit was another factor that determined the process of diversification as households with easy access were unlikely to diversify away from agriculture. Female-headed households which generally earned less income from agricultural production had limited opportunities to supplement income apart from engaging in casual labour. Another study by Chibwana et.al (2012) demonstrates that while there is a positive correlation between participation in the input subsidy programme and the amount of land used for planting maize and tobacco, empirical findings showed less diversification. This is due to the fact that farmers allocated more land to maize and tobacco and less land to crops such as groundnuts and soya beans.

Recent agricultural policy developments in post-independence Malawi demonstrate that while so many policies with good intentions have been formulated and implemented, the challenge of achieving food security has remained elusive. The country continues to experience frequent food shortages which are largely due to anthropogenic factors. The failure of intensification programmes to increase food production and food security remains as critical now as it was twenty years ago.

Discussion of Findings

This section presents key findings from the study by looking at issues of intensification and gender, crop diversification, and gender dynamics at household level. It shows how these ideas relate to specific aspects of the empirical data from the study.

Intensification and Gender

Agriculture remains the main livelihood activity for farmers in the four regions. Farmers have been using the same fields for a long period and growing the same kind of crops, dominated by maize. There are relatively few new areas where farmers could get additional farming land if they wanted to expand. Instead, further fragmentation is taking place where the same units of land are being shared among members of households. It was common for a household to have more than one plot at different locations. Some of the farmers who had slightly larger plots in Afrint II were found to have reduced plot sizes by the time of Afrint III. When asked to explain, the majority indicated that they had given out part of the land to their grown-up children. This was also confirmed by key informants both in the villages and among agricultural extension staff.

The farming land is still owned and inherited through the maternal side of the family, but in the case of married couples the farms are usually managed by men. In the quantitative study sample for Afrint III, 68.7 per cent were male farm managers while 31.3 per cent were female farm managers. Each household cultivates the family gardens and the produce is almost wholly used for home consumption or for sale if there is some surplus.

Most farmers depend on rain-fed agriculture although in some areas *dimba* (stream-bank gardens and low-lying areas using residual moisture during the dry season) fields are increasingly being utilized. The rainy season, which usually starts around November and ends in March or April of the following year, is the busiest and most active period for farming communities. *Dimba* fields have become important means of supplementing food requirements for households but also as sources of income. Empirical studies by Chinsinga (2007), Peters and Kambewa (2007), and Peters (2004a) also underscored the importance of *dimba* cultivation to household food security. Commonly grown crops during the winter season include maize, beans, tomatoes, and cabbages. Farmers who cultivate more than once a year have reported an increase in income levels as well as a general improvement in their livelihoods.

Estimated mean landholding that was put to the production of maize was, across the three rounds of surveys, less than 1 ha (Table 7.1). Holdings cultivated by households with female farm managers are slightly smaller than holdings cultivated by households with male farm managers and the differences are not statistically different. The household average yield of maize in the first round was no more than 4,668 kg/ha. This average has, over the years of the study, decreased to 1,092 kg/ha during round II and increased slightly in Afrint II to 1,248 kg/ha.[2] These findings are consistent with results from others

[2] This figure differs slightly from the figures presented in Chapter 2 (Table 2.5). The reason for this is that this chapter uses data only for the year of the survey, whereas the data in Chapter 2 are based on three-year annual averages.

Table 7.1. Mean area under maize cultivation (ha) and maize yield (kg/ha) by sex of farm manager

	Afrint I				Afrint II				Afrint III			
	Male	N	Female	N	Male	N	Female	N	Male	N	Female	N
Mean area under maize cultivation	0.15	167	0.13	136	0.88	222	0.61	171	0.87	275	0.65	125
Maize yield	4,882	163	4,406	133	1,096	222	1,087	171	1,257	197	1,235	140

Note: a. N equals the number of respondents who answered the questions in that category.

studies where the average yield in 2003/4 was 4,823 kg/ha, while between 1997/8 and 2002/3 it ranged between 1,274 and 1,309 kg/ha (RATES 2003: 10). Just as with average landholding sizes devoted to the growing of maize, the mean maize yield among households with female farm managers were in all rounds significantly less than the averages among households with male farm managers. Across the three seasons, the mean difference in production between households with female managers and those with male managers were not statistically significant across the three periods, but yield was slightly in favour of households with male farm managers.

Information and Technology

Apart from the amount of land that is devoted to a crop, the variety of seeds that a farmer uses matters as well. Our findings reveal that in the four districts under study, about 50 per cent of the households used traditional maize seed in the farming and this is consistent across the three rounds of the study. The other half used hybrid seed, and very few used open pollinated varieties or composite seeds. There are no significant differences between households with male farm managers and those with female farm managers. The widespread use of traditional seed varieties could be because of lack of resources to acquire these, but it could also be a result of limited knowledge regarding the value of planting improved seed varieties due to lack of extension services. Our study revealed that more than 50 per cent of the respondents had never received extension advice in the three rounds of the survey, and the percentage is as high as over 90 per cent in round III. Slightly more respondents from households with female farm managers had never had any extension advice than respondents from households that had male farm managers.

Another factor is the availability of fertilizer, particularly through the agricultural input subsidy programme, which has in some ways improved productivity. In principle, it is meant for needy members in the villages and especially the elderly, widows, and female-headed households. All such needy people are registered and they are supposed to get inputs. But there have been many operational challenges with the implementation of this programme. Sometimes village heads decide to distribute inputs to all residents in the village disregarding their socio-economic status. In other cases, the coupons are not enough, so that the needy people get only a pail and the rest may not get anything at all. The local criteria, however, means that all fertilizer is shared widely so that it is used by more people. This subsidy programme also encouraged the splitting of villages partly to increase chances of benefitting from the inputs but also to avoid conflicts in the village. But the criteria for registering beneficiaries have been changing every year depending on the amount of resources available at national level. At village level, generally, the typical amount of fertilizer available and the number of beneficiaries have been declining over time. This could be responsible for the decline in yields as seen earlier, and this is regardless of the gender of the farm manager since levels of poverty in the villages tend to be the same.

Crop Diversification

As alluded to earlier, over the years maize was the main crop that every household was supposed to grow to ensure that the household was food secure. In the face of climate change, dwindling land sizes, and reduction in the productivity of maize there has been a push from both the state and non-state actors for households to diversify the crops that they grow. Results of this study show that people have slowly taken to the idea as most households grow several crops and not just maize. When presented with demand, farmers tend to respond positively as is the case with increased production of groundnuts, soya, and the hybrid varieties of potatoes and maize. Although most of these crops were being grown in some areas in the past, respondents interviewed during the qualitative fieldwork indicated that production was generally low (women's focus group discussion in Chikwanje, 24 November 2012 and men's focus group discussion in Nkhwangwa, 26 November 2012). They further pointed out that low prices and absence of markets in the area were some of the reasons for low production. The interest in diversifying crop production may be attributed to farmers' sensitivity to market opportunities as well as a strategy for coping with climate variability. Table 7.2 demonstrates that more than 50 per cent of the sample over the three rounds grew more than four different types of crops.

Analysis of the number of crops that people grew by gender of the farm manager shows little difference in households where farm managers were men and where farm managers were women. Only in round III were differences significant at the 10 per cent level, with female-managed farms (FMFs) on

Table 7.2. Share of households that grew a certain number of crops by sex of farm manager, Afrint I to Afrint III

Number of crops	Afrint I				Afrint II				Afrint III			
	Male	N	Female	N	Male	N	Female	N	Male	N	Female	N
1	1.8	165	3.1	130	7.1	177	7.0	141	5.8	276	5.7	124
2	9.7		10.0		22.1		20.3		9.8		6.3	
3	13.3		13.1		16.8		18.0		7.6		11.1	
4	15.6		21.5		22.6		23.3		9.1		16.7	
5	19.0		20.0		14.6		12.8		12.0		15.1	
6	16.3		10.0		7.1		8.7		11.6		13.5	
7	7.9		13.1		8.4		7.0		13.4		11.1	
8	6.7		5.4		0.9		2.9		13.8		12.7	
9	5.5		2.3		0.4		0.0		9.4		1.6	
10	1.8		1.5		0.0		0.0		4.0		3.2	
11	1.8		0.0		0.0		0.0		1.4		2.4	
12	1.2		0.0		0.0		0.0		2.2		0.8	

Note: a. N equals the number of respondents who answered the questions in that category.

average being slightly less diversified. This means that all households regardless of the gender of the farm manager are diversifying probably because they are all subjected to the same kinds of environmental and economic conditions. There are slightly fewer females diversifying probably because of the need to produce food for their households.

In this study, we classified maize, rice, and sorghum as the major crops that people grew. When we consider a combination of the three, most households in that respect grow one of the three crops (Table 7.3). This is true of the data from rounds I, II, and III. In all rounds, more households that had female farm managers than those with male farm managers grew one major crop. Thus, in terms of major crops grown there was little diversification, especially among households with female farm managers in the last two rounds of the study. This implies that when it comes to issues of diversification, the gender of the farm manager does not matter. Since what we classified as major crops are largely subsistence crops, there is one crop that is dominant and that is maize. Messages of diversifying into other food crops may thus not be working.

The predominant crop grown by most people was maize. More than 95 per cent of the households interviewed across the three rounds grew maize. Less than 25 per cent grew rice and less than 2 per cent grew sorghum. Sorghum is one of the crops that is being slowly left out and replaced by other crops, both for food and non-food purposes. Differences between households with female farm managers and those with male farm managers were significant at the 5 per cent level for rounds I and II, but the differences were very small for both years and therefore need to be interpreted with caution.

Table 7.3. Number of major crops grown by sex of farm manager

Number of crops	Afrint I						Afrint II						Afrint III			
	Male	N	Female	N	Diff.	Sig	Male	N	Female	N	Diff.	Sig	Male	N	Female	N
1 or none	76.2	168	76.5	136	−0.003	**	76.1	226	80.2	172	−0.041	**	76.1	276	80.2	126
More than 1	23.8		23.5				23.9		19.8				23.9		19.8	

Note: a. N equals the number of respondents who answered the questions in that category.

Other Crops

Apart from what are regarded as staple food crops, households in the study sample also grew other food crops, including vegetables, beans, sweet potatoes, groundnuts, peas, plantains, cassava, Irish potatoes, and millet (Table 7.4). Except for beans, peas, and cassava, a higher proportion of farms with male managers grew these crops than those with female managers. These are crops that are usually sold for cash. This may suggest a gender dimension that male farm managers grow crops that are usually sold while crops that are predominantly grown for consumption are the responsibility of women. In most cases, more than a third of households that grew other food crops, including vegetables, sold them. This was the case regardless of whether the household was headed by a man or a woman. The differences between the households were rather small. In round one though, the differences in the share of female- and male-headed households that grew several of the crops were statistically significant. Likewise, in Afrint II differences were statistically significant for several of the crops. In the final data collection round, however, it was only households that grew sweet potatoes and those that grew millet that recorded significant differences between male and female farm managers.

Further investigations into crops that people grew showed that very few households grew non-food cash crops and these were mainly cotton, sugar cane, and tobacco, which were dominated by households that had male farm managers (Table 7.5). Thus, when it comes to growing high-value crops, households with female farm managers are in a minority. In round I, the only statistically significant difference between proportions male and female farm managers were those growing tobacco. In round II it was those growing tobacco and those growing sugar cane. In round III, significant differences were recorded for all the three crops.

Table 7.4. Share of households that produced other food crops, by sex of farm manager, Afrint I, II, and III

	Afrint I						Afrint II						Afrint III					
	Male	N	Female	N	Diff.	Sig.	Male	N	Female	N	Diff.	Sig.	Male	N	Female	N	Diff.	Sig.
Vegetables	71.4	168	60.4	134	11.0	**	40.5	195	40.1	152	0.4		68.1	276	69.8	126	−1.7	
Beans	60.1	168	58.5	135	1.6		68.6	194	71.4	154	−2.8		55.1	276	60.3	126	−5.2	
Sweet potatoes	61.9	168	57.0	135	4.9		46.7	195	47.1	153	−0.4		58.0	276	49.2	126	8.8	*
Groundnuts	42.9	168	33.3	135	9.6	*	47.2	195	28.8	153	18.4	***	50.0	276	43.7	126	6.3	
Peas	20.2	168	32.6	135	−12.4	**	18.6	194	35.7	154	−17.1	***	36.6	276	38.1	126	−1.5	
Plantains/ bananas	35.7	168	44.4	135	−8.7	*	23.6	195	31.2	154	−7.6	*	38.4	276	32.5	126	5.9	
Cassava	12.4	226	25.0	172	−12.6	***	18.1	226	29.1	172	−11.0	***	27.2	276	31.0	126	−3.8	
Irish potatoes	27.4	168	12.6	135	14.8	***	24.9	193	14.4	153	10.5	**	20.7	276	15.9	126	4.8	
Millet	11.3	168	11.9	135	−0.6		14.0	193	13.0	154	1.0		13.8	276	21.4	126	−7.6	**

Note: a. N equals the number of respondents who answered the questions in that category.

Table 7.5. Proportion of households growing non-food cash crops by sex of farm manager

	Afrint I						Afrint II						Afrint III					
	Male	N	Female	N	Diff.	Sig.	Male	N	Female	N	Diff.	Sig.	Male	N	Female	N	Diff.	Sig.
Cotton	1.2	168	0.0	136	1.2		3.6	136	2.9	170	0.7		9.1	276	4.0	126	5.1	*
Sugar cane	14.9	168	15.4	136	−0.5		7.6	136	2.4	169	5.2	**	14.9	276	5.6	126	9.3	***
Tobacco	20.8	168	4.4	136	16.4	***	14.7	136	1.8	170	12.9	***	14.1	276	2.4	126	11.7	***

Note: a. N equals the number of respondents who answered the questions in that category.

Crop Diversification Initiatives and Livelihood Changes

Commercialization of non-staple food crops was introduced by either NGOs or special production and marketing programmes. In Ntchisi, several organizations have been actively supporting commercial production of groundnuts, soya, and maize. These include World Vision and the National Association of Smallholder Farmers in Malawi, which provide groundnut seed. Farmers are expected to pay back at the end of the harvesting season double the amount of seed they received. In Thiwi Lifidzi, the Foundation for International Community Assistance, a micro-finance organization, has been organizing farmers to establish credit groups. Women in particular have benefited a lot from the Credit Union of Malawi Organisation, another NGO providing credit facilities, which facilitated the opening of a bank account from which farmers get loans to buy inputs. They have also been quite instrumental in securing markets for produce. Concern Universal used to support farmers through implementation of the APIP. A few donors also supported projects such as the Rural Economic Enhancement Project, which linked potato and groundnut farmers to markets like Universal Industries Limited. Similarly, the International Potato Centre project has been supporting farmers to increase production and marketing of potatoes in the two districts (key informant interview with District Agricultural Development Officer, Ntchisi, 23 November 2012; and District Agricultural Development Officer, Dedza District, Thiwi Lifidzi, 20 November 2012). Although the NGOs and micro-finance institutions provide an important service to the rural farming communities, the challenge is that they tend to operate in localized areas. Hence the results may be difficult to replicate or even scale up beyond the designated areas of intervention. In addition, regardless of the success of the projects, most donor-supported initiatives tend to fall off when such support phases out (Shivji 2006).

Some noticeable improvements in the livelihoods since the previous rounds of the survey were mentioned in the qualitative interviews. In many of the villages, and particularly in Khasu, there are more houses with iron sheets, the number of people keeping cattle has gone up, and in general more people can send their children to school. These changes can be attributed to increased income from the sale of crops. Farm-gate prices have also increased; for instance, a 200 kg bag of potatoes was selling at MK12,000—13,000 if sold to a vendor in the village—but the price went up to MK15,000 if sold at Chimbiya market. Farmers considered this to be a reasonably high increase from what they used to get in 2007/8 and this encouraged them to grow more potatoes (key informant interview with lead farmer at Khasu village, 21 November 2012).

One of the greatest challenges for farmers remains the issue of marketing. For example, while a market point exists at Chimbiya which is situated some twenty kilometres away from Khasu, few farmers can afford to transport their commodities and be assured of getting reasonable prices in good time. In the villages of Chikwanje and Nkhwangwa, Ntchisi *boma* (town), situated at a distance of between twenty and thirty kilometres, is the main market outlet for farmers produce. This situation has created space for the entry of middlemen, commonly known as vendors. Since the economic liberalization of the 1990s, middlemen from urban areas have penetrated many rural farming areas where they offer highly variable prices. In some cases they offer competitive prices, and in other cases they exploit farmers with rates that are below production cost. Farmers complained that vendors exploit them in many ways, such as through tying up of weighing scales, bargaining, or mixing up the figures.

Contribution of Farming to Household Income

Farming appears to be the backbone of rural livelihoods in Malawi, as most income was drawn from farming-related activities. Analysis of variance in round I shows that differences between households with male farm managers and those with female farm managers in terms of incomes from sale of non-food cash crops, non-farm salaried employment, and remittances were statistically significant, but remittances were skewed towards female farm managers. In round II the significant differences had increased and were now found in all farm-based sources of income that involved the sale of produce (staple crops, other food crops, non-food cash crops, and animal products). With the exception of the sale of animal produce, these differences remained in round III. Diversifying into these types of crops therefore seems to pay dividends, but has also led to widening gender gaps in commercialization.

Gender Dynamics at Household Level

The two rounds of this study revealed that maize and vegetables are the most predominant crops that people grew. Earlier studies in Malawi (Dorward et al. 2008) demonstrate that the ratio of maize growers in the country's four regions varied between 93 per cent and 99 per cent. This is consistent with the findings of this study as 98.7 per cent and 98.8 per cent of households with male and female managers, respectively, grew the crop in round II. The corresponding proportions in round III were 99.6 per cent and 99.2 per cent, respectively. Rice, on the other hand, was only grown in one of the regions covered by the study and sorghum was hardly grown at all. Patterns of production and growth are really based on the geographical location of the households. Rice could only be grown in one region because that is where conditions were ideal for production of this crop. What is striking about other food crops though is that these are grown everywhere and, in all the Rural Development Programmes, more households with female farm managers grew crops that were used as relish or used with relish as condiments while households with male farm managers grew crops that could be sold. This indicates that for households with female farm managers, consumption was given more priority than cash generation.

Most households grew maize over one season but intercropped with other crops basically to fully utilize the small land holdings and to take advantage of some of the extension messages. Most people grew either hybrid or local maize and very few grew the improved varieties. Data from qualitative research indicated that this was the case because improved varieties are labour and input demanding and difficult to store after harvest (men's focus group discussion, Nkhwangwa, 26 November 2012). Production has generally increased over the years but some households are still food insecure. This is not very different from the situation in other countries in the Southern African region (Sitko and Jayne 2014a). As Chirwa et.al. (2008) show, maize production had, over the past four decades, been growing in real terms but this growth has failed to keep pace with the population growth, a fact that has led to some commentators advocating for the idea of feeding the African population through increased imports of maize. Households with female farm managers in our sample are thus worse off, and over the years they consistently produced about half of their male farm manager counterparts. Holden and Lunduka (2013), in a study of who benefited from Malawi's targeted input subsidy programme, also noted that households with female farm managers were more likely to be net buyers as they produced significantly smaller amounts of maize than households with male farm managers. Whiteside (2000) points out that a disproportionate number of households with female farm managers in Malawi can be described as poor or very poor. He also argues that micro-level studies have shown that households with female farm managers are particularly labour constrained, a fact that makes it difficult for them to take advantage of off-farm employment.

Some of the maize that is grown is sold, most of it to traders from the village and briefcase traders (vendors) from outside the village. Results show that the price of maize over the years had got better (in particular, 14.3 per cent of the respondents in round III indicated that prices had got worse, while 78.6 per cent indicated that prices had improved), but only 40 per cent of the sample indicated that the volume of maize that they sold had increased. This is probably because the volume produced had also diminished because of the decrease in land sizes.

Conclusion

This chapter started by raising three research questions regarding gender dynamics in agricultural intensification in the Afrint research areas of Malawi. It has been established that the desire to achieve food security has largely been responsible for propelling Malawi to adopt and implement agricultural intensification. But such intensification has not been systematic as policy changes have generally not been well coordinated. Many of the policy interventions were designed to benefit women and other vulnerable groups through an increase in farm income and food security. It has also been demonstrated that the declining land sizes and the provision of market opportunities have encouraged the process of crop diversification. Although government has in most cases formulated policies on diversification, implementation has until recently not been aggressively promoted. It is mostly NGOs and some donor-supported projects that have taken the lead, as confirmed by empirical evidence from our research sites. These organizations are actively organizing women to form groups so that they can access credit and inputs for effective production and marketing of their crops. NGOs are usually concentrated in small areas, but the initiatives would be of more benefit if they were replicated and up-scaled across the country.

8

Policies or Prices? A Gendered Analysis of Drivers of Maize Production in Malawi and Zambia, 2002–13

Martin Prowse and Ellen Hillbom

Introduction

Ever since its introduction some 500 years ago, maize has played an important role in processes of long-term economic, political, and social change in Africa. It has to a large extent supplanted the continent's own historical grain crops such as sorghum and millet, and has supported increasing populations (McCann 2005). White maize dominates food consumption in Eastern and Southern Africa and accounts for more than half the calories people consume. Although many rural households in Malawi and Zambia are net food purchasers, almost all of them produce maize for consumption. While demand for maize is expected to stay high in the foreseeable future (Byerlee et al. 2006, Jayne et al. 2010), the deepening of agricultural markets (due to changing food preferences, urbanization, and population growth) as well as better local, national, and regional labour markets, provide contemporary smallholders with a growing range of opportunities for diversification of agricultural production and livelihood strategies. Under such circumstances it is pertinent to ask to what extent and how smallholders are motivated to improve productivity.

In this chapter, we pursue this line of questioning by using Afrint data from Malawi and Zambia during the period 2002 to 2013. The comparison is instructive as both neighbours are landlocked and share some linguistic, matrilineal, and colonial similarities. However, they also differ in terms of population densities, landholding sizes, industrial structure, and policies encouraging smallholder production. The chapter applies an explicit gender lens to examine if we can detect differences in maize farming by comparing farms managed by men and women. We locate the study within each country's historical context of post-independence government agricultural policies with a particular focus on price stabilization and subsidy schemes. As the Afrint I, II, and III rounds of data correspond with the most recent rise and fall in global price on maize, we also explicitly investigate the extent to which the global commodity boom of the early 2000s influenced smallholder production. Further, we utilize qualitative data collected in Malawi and Zambia in 2012 and 2016 in the form of semi-structured interviews with key informants and household respondents as well as focus group discussions to elaborate on the results obtained from the Afrint panel data.

The structure of the chapter is as follows. In the next section we examine the basic trends in maize production and productivity in Malawi and Zambia from Afrint I to III. We then contextualize long-term government agricultural strategies. Following this, we discuss the expected relationship between hikes in global maize prices and national producer prices. We then turn to the Afrint data presenting changes in maize production during 2002–13 across farms managed by men and women. Next, we conduct a production function to assess the relative contribution of household-level factors to productivity increases. Thereafter we turn to the qualitative data and the last section concludes.

Trends in Maize Production and Productivity in Malawi and Zambia, 2000–14

Figure 8.1 details recent country-level trends in maize production in Malawi and Zambia based on FAO data.[1] The data cover all farm sizes and all types of maize, although both Malawi and Zambia produce almost exclusively white maize. Production increased from 2.5 to 3.92 million metric tons ($r^2 = 0.711$) and from 1.04 to 3.35 million metric tons ($r^2 = 0.813$) in Malawi and Zambia, respectively. This trend is significant in both countries at the 1 per cent level using a one-sample t-test and corresponds to an annual growth rate of around 7 per cent and 11 per cent respectively. When we conduct a linear regression trend analysis for both countries it shows significance at the 1 per cent level (although the Zambian data do show signs of serial autocorrelation as measured by the Durbin–Watson test). In per capita terms, though, this absolute increase in maize production corresponds to stagnation in Malawi from 220 kg to 226 kg per capita compared to a rapid increase in Zambia from 103 kg to

[1] We recognize that these FAO data may be noisier than comparable Ministry of Agriculture data

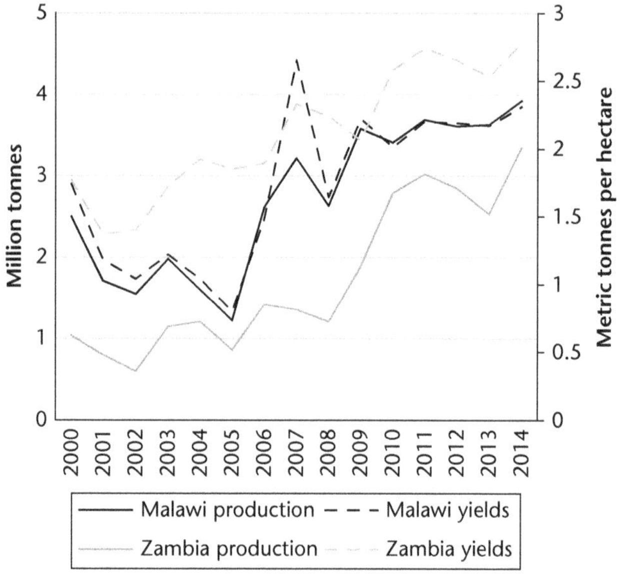

Figure 8.1 Trends in maize production (million tons) and yields (metric tons per hectare) in Malawi and Zambia, 2000–14.
Source: FAOSTAT, April 2016.

220 kg per capita.[2] Figure 8.1 also shows that yields, i.e. land productivity, have increased in both countries. In Malawi we see an increase from 1.74 to 2.31 tons per hectare ($r^2 = 0.51$) while the Zambian data show an increase from 1.77 to 2.78 tons per hectare ($r^2 = 0.85$). This corresponds to growth rates of 6 per cent and 5 per cent respectively. Both one-sample t-tests and linear regression trend analysis show significance at the 1 per cent level.

While capturing changes in production and productivity, these numbers do not, however, disclose which segments of maize producers are driving national trends: large- or small-scale farmers. Later in the chapter we return to this issue as we investigate to what extent and through which mechanisms smallholders in Afrint villages are part of the national trends.

State Intervention for Agricultural Development

Agricultural policies in Malawi after independence in 1964 initially focused on promoting smallholder production on customary land (Kydd 1984, Mkandawire

[2] Population estimates from the World Development Indicators and World Population Review for 2000 and 2014, respectively.

and Trust 1999). Part of this strategy was the reform of the agricultural extension and marketing services provided by the Farmers' Marketing Board. The state's increased control over every aspect of agricultural production and marketing during the 1960s, and the harsh penalties levied if these regulations were contravened, prevented both settlers and African intermediaries from competing with the Farmers' Marketing Board in the marketing of food and cash crops (Ng'ong'ola 1986). Due partly to a limited supply response from smallholder agriculture, and facing a balance of payments crisis (Baker 1962, Thomas 1975), the early 1970s saw a switch in government policy away from the smallholder sector towards estates as a source of agricultural growth (McCracken 1984). The smallholder sub-sector was now viewed as a source of labour, as well as playing a role in the construction of the nation through providing maize for both rural and urban populations. This was achieved through the parastatal Agricultural Development and Marketing Corporation (ADMARC) offering fertilizer subsidies and high pan-territorial and pan-seasonal producer prices for maize. Meanwhile, consumer prices for maize were also subsidized. This state-led model worked relatively well until the end of the 1970s when decline in terms of trade, increases in international interest rates, drought, and the disruption of export routes exposed the frailties of the dualistic agricultural development strategy (Ng'ong'ola 1986, Harrigan 2001, Van Donge 2002).

In response to the economic crisis, the government opened negotiations with the International Monetary Fund and the World Bank, and structural adjustment programmes were introduced. During the 1980s, subsidies for maize seeds and fertilizers were phased out, and the producer price of maize was reduced relative to export crops (Lele 1990, Kherallah and Govindan 1999, Harrigan 2001). Harrigan (2001) argues that the consequence of increasing producer prices for export crops, while holding ADMARC maize prices stable, was production switching from maize to fire-cured tobacco, groundnuts, and cotton. This reduction in maize hectarage was exacerbated by the removal of the fertilizer subsidy programme which reduced the profitability of maize, especially high-yielding varieties. These factors, along with the financial collapse of ADMARC, reduced maize production and precipitated a national food crisis in 1987, causing government to reintroduce fertilizer subsidies and increase maize producer prices (Harrigan 2005).

In the late 1990s there was a 'policy vacuum' regarding food production in Malawi. Government and donors implemented a piecemeal reactive series of measures (Harrigan 2003). The 1996/7 and 1997/8 agricultural seasons saw a dramatic decline in maize production. From the 1998/9 agricultural season a Starter Pack scheme reversed this trend, contributing to bumper maize harvests in 1998/9 and 1999/2000. In subsequent years, Starter Pack was accused of crowding out the private sector, fostering smallholder dependency on hand outs, and being overly bureaucratic (Harrigan 2005, Levy 2005), and it evolved into a 'targeted' social safety net in the form of the Targeted Input Programme (TIP). Two events which shook policymakers into action were the regional food security crises of 2001/2 and 2005/6. Policymakers responded by implementing input subsidies to boost maize production and productivity. The creation of Malawi's Farm Input Subsidy Programme in the 2005/6 season more than doubled maize production from 1.2 million to 2.6 million metric tons. Between the 2006/7 and 2013/14 seasons, production only dropped below 3 million metric tons in 2007/8 (Chirwa and Dorward 2013, Arndt et al. 2016).

Meanwhile, in Zambia the colonial era saw the establishment of the mining sector as the main export earner, and the rural areas as providers of labour and food. Preferential policies were developed with the purpose of creating a smaller group of commercialized maize farmers, initially settlers but later including Africans (Gann 1969). After independence in 1964, the government chose to subsidize agricultural inputs to continue stimulating maize production by the commercial sector to feed its (relatively, in African terms) large share of urban population, but also to encourage smallholder production. While large-scale farmers demonstrated significant progress based on mechanization, subsidies, new hybrid varieties, and farming on marginal lands, the smallholder sector also improved. Overall, from the early 1960s to the late 1980s, Zambian farmers increased their maize production by 400 per cent (McCann 2005: 162). However, the economic crisis in the 1980s and early 1990s resulted in structural adjustment reforms which required a significant reduction of subsidies. The lack of extension and credit for hybrid maize production as well as market prices for fertilizer caused many smallholders to revert to indigenous crops such as sorghum, millet, beans, and groundnuts (Jayne and Jones 1997: 166, Govereh and Jayne 2003, McCann 2005).

In 1995 the Food Reserve Agency (FRA) was established as a parastatal strategic food reserve/marketing board with a mission to buy maize at a pan-territorial price exceeding wholesale prices in major maize-producing areas. The FRA's price policies included, for example, offering farmers above-market prices, subsidizing prices to select large-scale millers, and exporting regionally (even at prices below the FRA purchasing price). FRA activities have both made maize prices more stable and raised them above average market prices, by as much as 17–19 per cent for the 2003–8 period. Between the 2004/5 and 2010/11 growing seasons the FRA purchased 30–86 per cent of maize marketed by smallholders, costing the government roughly 25 per cent of the annual total agricultural sector budget (Mason and Myers 2013). The FRA's interventions in output markets and price control represent active state policies that set Zambia and Malawi apart. While in theory all smallholders can sell maize to the FRA, unreliability and delays of payments by the parastatal entice those in urgent need of cash to seek alternatives buyers of their produce, often selling well below the prices set by the government. Further, many cooperatives profit from the opportunities represented by the price gap, turning into maize-trading, rather than maize-producing, groups. Finally, the use of the FRA's pricing mechanisms to boost incomes of certain cooperatives and private agents has contributed to the considerable growth of medium-scale and large-scale commercially oriented farming, and increased the gap between them and the small-scale subsistence farmers (Sitko and Jayne 2014b, Andersson Djurfeldt and Hillbom 2016).

As a complement to the FRA, the Fertilizer Credit Program was created in 1997/8 with a small subsidy attached (Mason et al. 2012). In the 2002/3 season, large-scale fertilizer subsidies were reintroduced. During the three years the programme ran, on average 29,000 metric tons of fertilizers were distributed annually, particularly to the major maize-producing regions in the Central, Eastern, and Southern provinces. However, repayment rates were poor, and the next programme, the Fertilizer Support Program, was designed as a cash-only input subsidy programme targeting selected beneficiary farmers. Annual volumes were slightly more than double compared to the previous programme, and beneficiaries were somewhat more evenly distributed over the country. In 2009/10 the programme was renamed the Farmers Input Support Program (FISP), and it was still running at the time of Afrint III. The volume and number of recipients has been substantially increased from 48,000 metric tons of fertilizers and 120,000 farmers in 2002/3 to 180,000 metric tons and 900,000 farmers ten years later (Ricker-Gilbert et al. 2013). FISP is open to households with more than 0.5 ha of land and it is delivered through registered farmer groups such as cooperatives. These principles mean that the programme excludes 15–20 per cent of the poorest rural households in the country (Sitko and Jayne 2014b).

The fertilizer subsidy schemes in both Malawi and Zambia have been contentious (see Chinsinga and O'Brien 2008, Chirwa and Dorward 2013, Sitko and Jayne 2014b, Arndt et al. 2016, Ricker-Gilbert and Jayne 2017). The challenges for implementation of fertilizer subsidies in both countries stems from two forms of market failure. First, the slow process of rolling back the state within agricultural input markets created an uncertain policy environment, which resulted in thin fertilizer markets in both countries characterized by lack of authentic competition and penetration into all regions. In addition to the overriding political imperative of assuring national-level food security in the context of uncertain climatic conditions, imperfections within fertilizer supply have exacerbated maize price volatility (including speculation by, e.g., maize-trading cooperatives in Zambia), placing pressure on the social contract in both countries. The second form of market failure has been in credit markets. Since the late 1990s, financial institutions have not been able to offer credit to smallholders to intensify maize production on a sustainable basis due to them consistently prioritizing immediate consumption needs over longer-term repayment considerations.

Reflecting the focus of this chapter, gender has increasingly been integrated within national policy frameworks in both countries. Within Malawi, the overarching Agricultural Sector-Wide Approach (ASWAp) aims to ensure that half of participants in programmes, such as the FISP and the Green Belt initiative on irrigation, are women. The ASWAp recognizes that women-headed households have worse access to land, labour, and credit than male-headed counterparts and, similar to the national gender policy, increase women's control over agricultural resources and technologies. In Zambia, gender within agriculture is conspicuous by its absence: the Sixth National Development Plan relegates gender to being 'mainstreamed' and delegates coordination to the Ministry of Gender and Child Development. Gender also remains marginal within the National Agricultural Investment Plan (NAIP): it is seen as a cross-cutting issue. The main positive statement in the NAIP is the goal that 30 per cent of targeted beneficiaries are to be women.

Global and National Maize Prices, 2002–14

After around twenty-five years of decline, internationally traded food commodity prices started to increase sharply from January 2005 to June 2008 (Mitchell 2008, Piesse and Thirtle 2009). Being both a significant staple food crop in much of the developing world and having multiple uses including food, feed, and fuel, maize was at the core of the food commodity price boom, with prices tripling (Rosegrant 2008, Piesse and Thirtle 2009). After their peak in 2008, cereal prices declined by a quarter, only to increase rapidly again in 2011/12 before subsiding to below 2007 levels in early 2015 (World Bank 2015). Maize prices broadly reflect these general trends, showing a 20 per cent decline up to the year ending May 2015 (World Bank 2015). Biofuels were initially seen as a window of opportunity for African smallholders and a route to pro-poor agricultural growth (Von Braun and Pachauri 2006, Peskett et al. 2007). However, it became increasingly clear that the biofuel boom (itself driven by high oil prices) was driving higher food prices as broader crops, e.g. maize, were close substitutes in consumption and production (Mitchell 2008, Rosegrant 2008).

The percolation of increasing prices on international markets to national producer prices benefitting the commercialized smallholder population in countries such as Malawi and Zambia should not be taken for granted. The development challenge inherent in occupying a relatively remote geographical location lies in high transportation costs, making it difficult to realize gains from trade. These challenges are particularly severe for countries with underdeveloped technology and infrastructure and a reliance on passage through sovereign transit countries (Sachs and Warner 1997, Mitchell 2008: 5). Not surprisingly, there is strong evidence that a country's level of infrastructure development, i.e. transportation costs, is associated with its level of agricultural productivity (Antle 1983, Gollin and Rogerson 2014). Further, when a high import/export parity wedge and underdeveloped domestic markets are combined with high domestic production variability, the impact of domestic shocks such as droughts is increased. Consequently, domestic price instability tends to be highest in landlocked African countries (Byerlee et al. 2006: 277). Meanwhile, regional trade, in combination with good transportation infrastructure between countries, has the potential to expand the size of the market and reduce price instability (Dorosh et al. 2009).

In Figure 8.2 we plot the development of national producer prices for white maize in Malawi and Zambia for 2002–14. The data show that the basic trends to a significant degree conform with the global hike in maize prices, though there are variations. First, the sharp increase in prices is delayed and not recorded until 2007. Second, while prices are high in 2008 the final peak is in 2009. Third, as is the case for global prices, national prices turn down, but

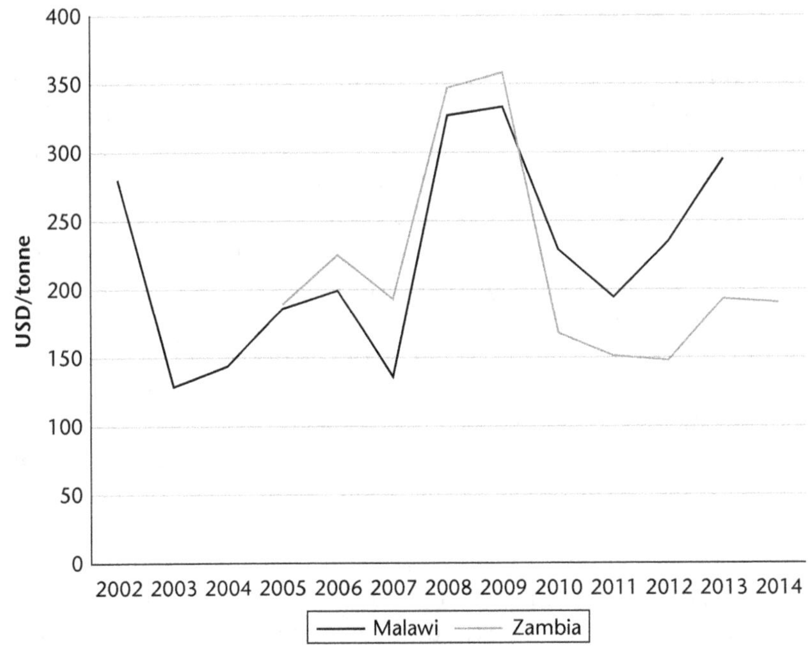

Figure 8.2 National white maize producer prices in USD/ton in Malawi and Zambia, 2002–14.
Source: FAOSTAT, April 2016.

increase again somewhat in 2011. In sum, during our period of investigation, 2002–13, national producer prices for white maize are about the same at the beginning and at the end of the period in both Malawi and Zambia, but in between there is a price hike stretching from 2007 to 2009/10.

While the commodity boom left net developing-country consumers living in poverty vulnerable to increasing food prices, it also created opportunities for primary sector-based economies to profit by increasing their own production (Helbling et al. 2008). In the rearticulation of smallholder-led agricultural development around the turn of the millennium, it was argued that smallholders broadly respond to market incentives, including price movements (World Bank 2007). Hikes in food prices, including maize, could thereby be seen as constituting a window of opportunity for producers. Scholars have indeed found that higher profits from maize production caused African farmers generally to shift to maize cultivation (Rosegrant 2008, Ghosh 2010, Jayne et al. 2010), but we want to see to what extent the hike in national maize prices influenced Afrint smallholders' maize production in Malawi and Zambia specifically.

Descriptive Statistics

We build on the three rounds of Afrint quantitative household data, described in Chapter 1. As far as possible, we conduct consistent comparisons between households with men and women as farm managers, referred to as male-managed farms (MFMs) and female-managed farms (FMFs) respectively. The three waves of data show considerable mobility between these household categories. As we do not have a variable for sex of farm manager for Afrint I, we are forced to use the household headship variable as a proxy. Within the Malawian data we have 268 households in the panel covering both waves. Afrint I shows 195 male-headed households (73 per cent) and seventy-three women-headed households (27 per cent). In Afrint II, we see an increase in the balance of women with 118 households (44 per cent) containing FMFs compared to 150 MFMs (56 per cent). In Afrint III, we see a reduction again to ninety-nine FMFs (37 per cent) vs 169 MFMs (63 per cent). To ensure we are truly tracking the same households through time, we only use data from the 145 MFM households and ninety FMF households which display the same household categorization throughout both waves (all three cross-sections) of the dataset. In other words, we exclude the households which changed household status through the duration of the panel. The proportion of women managing farms in the Zambian panel of 276 households showed similar levels of mobility between categories, and again we only use the households which we are certain had the same gender of farm manager throughout the period of investigation.

Table 8.1 presents selected descriptive statistics of relevance for our investigation, i.e. trends in maize production, area under maize farming, yields, and use of fertilizers. It is based on true panel households across both waves in Malawi and Zambia, and shows results for all households as well as for male and female farm managers separately. Significance is assessed through paired-sample t-tests.

As expected, the application of fertilizer to maize has increased during 2002–13 for all farm categories. In both countries, men have increased the application of fertilizer to a much greater extent than women: in Zambia by 210 kg more, in Malawi by around 47 kg more. Overall, maize production has increased significantly during the period in question: by 22 per cent in Malawi and by a staggering 239 per cent in Zambia. This means that trends in national maize production that we noted earlier are not driven by large-scale farms only, rather the increase in production in the smallholder sector is higher than the national average, and significantly so in Zambia. While households in Zambia were producing around double their Malawian counterparts in Afrint I, by Afrint III this had grown to more than a factor of four. Increases in maize production are coming from farms managed by men, not women. In statistical terms, farms managed by women did not see any significant changes in maize production between Afrint I and Afrint III. However, yields have increased dramatically in both countries for farms managed by women. In Malawi, women increased yields by 26 per cent compared to 47 per cent for men. In Zambia, women increased yields by 51 per cent compared to men's increase of 84 per cent.

At the same time as maize production is increasing, MMFs in Malawi have reduced land allocated to maize from 0.95 to 0.88 ha, and FMFs in Malawi have reduced the area under maize by 0.19 ha. Meanwhile, the story is the opposite in Zambia, where men have increasing land under maize from 1.47 to 1.88 ha and women have increased land under maize, although to a slightly lesser extent (and the result is not significant). For Malawi these results resonate with the trends found at the national level and supports the conclusion that increased maize production is a process characterized by intensification. This is a land-scarce country, which is demonstrated by national numbers reporting 183 people per square km (World Bank 2016). The qualitative Afrint data show that the sampled sites in Malawi generally follow the national trends reporting high levels of population density. In Zambia, while the national trend pointed towards intensification, the Afrint data indicate that for the smallholder population increased maize production is instead a story of extensification. National population density levels are at low, at twenty-two people per square km (World Bank 2016), but the Afrint sites in Mazabuka and

Table 8.1. Maize area, production, and yield for all households with male and female farm managers in Malawi and Zambia, Afrint I–III

Afrint		Area under maize, 3-year average (ha)						NPK application to maize (kg)					
		I	III	I	III	I	III	I	III	I	III	I	III
		All		Male		Female		All		Male		Female	
Malawi	Mean	0.90	0.79	0.95	0.88	0.82	0.63	32.18	136.14	34.47	154.05	27.79	101.69
	No.	232	232	143	143	89	89	152	152	100	100	52	52
	S.D.	0.51	0.56	0.49	0.62	0.52	0.40	51.97	137.28	52.36	158.35	51.44	72.68
	Sig.	***				***		***		***		***	
Zambia	Mean	1.40	1.76	1.47	1.88	0.97	1.04	214.27	526.51	227.68	575.77	149.11	287.21
	No.	198	198	169	169	29	29	164	164	136	136	28	28
	S.D.	1.05	1.42	1.07	1.48	0.81	0.67	398.45	595.67	428.94	618.06	182.13	400.55
	Sig.	***		***				***		***		***	

Afrint		Maize production (kg)						Yields (kg per hectare), based on pruned area data					
		I	III	I	III	I	III	I	III	I	III	I	III
		All		Male		Female		All		Male		Female	
Malawi	Mean	745.43	906.38	801.72	1,098.25	652.95	591.17	879.35	1.225.69	903.08	1.326.77	840.95	1.062.02
	No.	222	222	138	138	84	84	220	220	136	136	84	84
	S.D.	645.31	1.034.23	600.81	1213.89	706.45	506.45	616.81	876.31	688.35	947.06	480.37	723.55
	Sig.	**		*				***		***		***	
Zambia	Mean	1.655.23	3.951.37	1.721.94	4.442.46	1.330.86	1.651.15	1143.97	1998.63	1158.75	2.125.41	1.072.64	1.618.68
	No.	170	170	141	141	29	29	169	169	140	140	29	29
	S.D.	2.219.82	5.247.31	2.322.05	5.606.03	1.627.97	1.554.33	716.42	1.196.95	717.90	1.238.32	717.43	873.70
	Sig.	***		***				***		***		***	

Table 8.2. To what extent do the trends in maize cultivation differ between households with male and female farm managers?

		Area under maize, 3-year average, change Afrint I–III (ha)		NPK application to maize, change Afrint I–III (kg)		Maize production, change Afrint I–III (kg)		Yields (kg/ha), using pruned area data	
		Male	Female	Male	Female	Male	Female	Male	Female
Malawi	No.	143	89	100	52	138	84	136	84
	Mean	−0.18	−0.07	119.58	73.90	296.52	−61.77	423.70	221.07
	S.D.	0.6	0.71	154.69	75.70	1263.51	822.66	1,078.32	779.41
	Sig.			**		**			
Zambia	No.	169	29	138	28	141	29	29	140
	Mean	0.41	0.07	348.09	138.11	2,702.53	320.29	546.04	966.66
	S.D.	1.51	1.01	638.99	393.72	5,486.70	2,182.36	1,115.69	1,407.92
	Sig.			*		**			

Mkushi show regional differences. While smallholders in Mkushi state that much of their arable land is left idle, in Mazabuka this is not the case. Nevertheless, it seems that Afrint smallholders have sufficient land to extend rather than intensify their farming. Table 8.2 confirms the findings from paired-sample t-tests and shows through independent sample t-tests that men significantly increased fertilizer application and maize production to a greater extent than women, but that changes in area and productivity between MMFs and FMFs show no significance.

Production Function on Malawian and Zambian Data

To further explore changes in production and productivity we now turn to assessing the relative importance of a number of household factors. We follow the conventional specification of a production function as described in Peterman et al. (2011), logging both sides of the equation. As our dependent variable is normally distributed, we use a logged ordinary least squares regression model. To cover the full period of the project we use Panel I and Panel II data. For each country, we pool data for households with male and female farm managers and include gender of farm manager as a dummy to see how it influences productivity. Second, we run the same model stepwise with gender of farm manager introduced in Block 2 to compare coefficients and how the inclusion of gender of farm manager alters the model.

In previous sections we hypothesized that government policies generally and fertilizer subsidies specifically, as well as national producer prices, could be expected to influence smallholder maize yields (the dependent variable). To capture government policies in the model we have included the use of NKP (fertilizers) and improved seeds in both countries. In addition, as the qualitative data for Zambia informed us that extension officers' promotion of specific conservation farming techniques, i.e. ripping, was deemed to have considerable positive effects on maize production, this variable was also included, but only for Zambia. Maize prices are complimented with two spatial variables: distance to market and village centre. Further, we look into smallholders' access to the two basic factors of production—land and labour. For land we include size of land used for maize production and for the quantity of labour we have number of able workers and hiring of temporary farm hands, so-called *ganyu* in Malawi. We also have more qualitative aspects of labour captured by age and years of schooling of farm manager. Finally, we capture the gender perspective as sex of farm manager.

We discussed earlier how the longitudinal study limited the sample size as we use data from true panel households. The pruning of continuous variables further reduces the number of cases available for the ordinary least squares regression model. Despite these limitations, we obtain significant results. Descriptive statistics for the Malawian model with relevant transformations of variables are shown in Table 8.3. Missing values for some continuous variables have been imputed using median or mean figures. Skewness and

Table 8.3. Descriptive statistics for Malawian production function

	N	Mean	Std. deviation
	statistic	statistic	statistic
Dependent			
Yields—pruned at 3SDs, logged	138	7.86	0.32
Independent			
Age of head of household, years—68, missing values imputed with mean	138	6.36	0.95
Years of schooling of farm manager—2013 variable logged, 7 values imputed with median	138	1.29	0.75
Estimate of trend in able workers Afrint I to III, plus constant, logged, 5 missing values imputed with median	138	1.64	0.36
Trend in maize area Afrint I to III—pruned at 3SDs, plus constant, logged, 7 missing values imputed with median	138	1.02	0.19
Ganyu nominal trend Afrint II to III, plus constant, logged, 2 missing value imputed with median	138	4.24	0.50
NPK trend Afrint I to III— pruned at 3SDs, plus constant logged	138	5.61	0.54
Started using improved seeds between Afrint I to III, dummy	138	0.39	0.49
Started hiring labour in between Afrint I to III, dummy	138	0.25	0.43
Distance to market, household level, logged, 2008	138	1.86	0.64
Distance to village centre, logged, 2013, missing values imputed	138	−1.07	1.23
Trend in maize prices in USD nominal from Afrint II to III, no outliers, plus constant, logged	138	2.74	0.30
Dummy for sex of farm manager consistent across 2008 and 2013 with 2001 household head	138	0.64	0.48

kurtosis are below accepted limits. Listwise, 138 units are available for the production function. Table 8.4 shows coefficients for the model alongside significance figures. Collinearity statistics are within limits with variance inflation factors of variables below 1.2. The distribution of residuals is normal. Overall, the model explains 39 per cent of the variance of change in maize yields.

Table 8.4 shows that five independent variables are significant. The trend in maize area shows a strong negative relationship (−0.39) with maize yields at

Table 8.4. Coefficients for Malawian production function

Model	Unstandardized coefficients		Standardized coefficients	Sig.
	B	Std. error	Beta	
(constant)	7.448	0.433		***
Age of head of household, years—68 missing values imputed with mean	−0.056	0.025	−0.163	**
Years of schooling of farm manager—2013 variable logged, 7 values imputed with median	−0.009	0.030	−0.020	
Estimate of trend in able workers Afrint I to III, plus constant, logged, 5 missing values imputed with median	−0.017	0.064	−0.019	
Trend in maize area Afrint I to III—pruned at 3SDs, plus constant, logged, 7 missing values imputed with median	−0.669	0.123	−0.386	***
Ganyu nominal trend Afrint II to III, plus constant, logged, 2 missing value imputed with median	0.006	0.045	0.009	
NPK trend Afrint I to III— pruned at 3SDs, plus constant logged	0.284	0.041	0.476	***
Started using improved seeds between Afrint I to III, dummy	−0.030	0.047	−0.046	
Started hiring labour in between Afrint I to III, dummy	0.038	0.051	0.050	
Distance to market, household level, logged, 2008	0.100	0.035	0.198	***
Distance to village centre, logged, 2013, missing values imputed	0.049	0.018	0.186	***
Trend in maize prices in USD nominal from Afrint II to III, no outliers, plus constant, logged	−0.098	0.074	−0.091	
Dummy for sex of farm manager consistent across 2008 and 2013 with 2001 household head	0.003	0.048	0.004	
Model summary				
Model	r	r^2	Adjusted r^2	Std. error of the estimate
1	0.668a	0.447	0.394	0.25162

Notes: a. B is the symbol used for an unstandardized ordinary least squares regression coefficient.

the 1 per cent level. Holding all else equal, the trend in fertilizer application shows a strong positive relationship (0.48) with the trend in maize yields at the 1 per cent level. Two spatial variables show weak positive and significant relationships with the trend in maize yields: distance to market (0.20) and distance to the village centre (0.19), both at the 1 per cent level. The direction of these relationships is counter-intuitive: one would expect households closer to density and markets to have higher levels of productivity. Finally, age shows a negative relationship (−0.16) at the 5 per cent level.

Table 8.4 also shows that counter to our initial assumptions regarding the impact of price incentives, the logged trend in maize price variable does not show any significance (although the construction of the indicator does not, at present, allow for a lagged supply response). Finally, the variable for gender of farm manager is insignificant. When the model is run stepwise with gender of farm manager inserted in Block 2, very marginal changes are detected, none of which alter significance levels of any variables. Thus, the data suggest that, despite a slightly different strategy towards maize cultivation during the period under question, households with female farm managers have been sharing in the productivity increases enjoyed by households with male farm managers, and are subject to the same broad influences in productivity.

Moving on, descriptive statistics for the Zambian model are shown in Table 8.5. In a similar fashion to the Malawian model, missing values have

Table 8.5. Descriptive statistics for Zambian production function

	N	Mean	Std. deviation
	statistic	statistic	statistic
Dependent			
Yields—pruned at 3SDs, plus constant, logged	151	7.98	0.46
Independent			
Age of head of household, years—square rooted, 9 missing values imputed with mean	151	6.70	1.03
Years of schooling of farm manager—2013 variable logged, 6 values imputed with mean	151	1.92	0.63
Estimate of trend in able workers Afrint I to III, plus constant, logged	151	2.58	0.29
Trend in maize area Afrint I to III, pruned at 3SDs, plus constant, logged, 7 missing values imputed with median	151	1.60	0.28
Ganyu nominal trend Afrint I to III, plus constant, logged	151	6.32	0.23
Started using conservation agriculture Afrint I to III, dummy	151	0.45	0.50
NPK trend Afrint I to III—pruned at 3SDs, plus constant, logged	151	7.44	0.22
Started using improved seeds between Afrint I and III, dummy	151	0.16	0.37
Started hiring in labour Afrint I to III, dummy	151	0.17	0.37
Distance to market, household level, logged, 2008	151	2.95	0.51
Distance to village centre, logged, 2013, missing values imputed	151	1.39	1.07
Trend in maize prices in USD nominal Afrint II to III, no outliers, plus constant, logged	151	2.84	0.20
Sex of farm manager consistent across Afrint II and III with Afrint I HH head, dummy	151	0.80	0.36

been imputed where appropriate, and skewness and kurtosis are below accepted limits. In this model, 151 units are available for the production function. Table 8.6 shows coefficients for the model alongside relevant further statistics. Collinearity statistics are within safe limits. The distribution of residuals is normal. Overall, the model explains a lower degree of variance than the Malawian model: only 22 per cent of the variance in maize yields.

Table 8.6 shows that similar to the Malawian model we find that fertilizer (0.30), distance to market (−0.20), and distance to the village centre (−0.19) show significance. However, and in contrast to Malawi, the coefficient signs for both spatial variables are negative, suggesting proximity to density and market channels is imparting a positive effect on yields. Moreover, we find

Table 8.6. Coefficients for Zambian production function

	Unstandardized coefficients		Standardized coefficients	Sig.
	B	Std. Error	Beta	
(Constant)	3.833	1.624		**
Age of head of household, years—square rooted, 9 missing values imputed with mean	0.014	0.035	0.032	
Years of schooling of farm manager—2013 variable logged, 6 values imputed with mean	−0.034	0.057	−0.047	
Estimate of trend in able workers Afrint I to III, plus constant, logged	0.269	0.121	0.174	**
Trend in maize area Afrint I to III, pruned at 3SDs, plus constant, logged, 7 missing values imputed with median	−0.040	0.133	−0.024	
Ganyu nominal trend Afrint I to III, plus constant, logged	−0.148	0.147	−0.075	
Started using conservation agriculture Afrint I to III, dummy	0.129	0.073	0.141	*
NPK trend Afrint I to III—pruned at 3SDs, plus constant, logged	0.618	0.168	0.303	***
Started using improved seeds between Afrint I and III, dummy	0.046	0.093	0.037	
Started hiring in labour Afrint I to III, dummy	0.094	0.092	0.076	
Distance to market, household level, logged, 2008	−0.129	0.071	−0.144	*
Distance to village centre, logged, 2013, missing values imputed	−0.075	0.034	−0.177	**
Trend in maize prices in USD nominal Afrint II to III, no outliers, plus constant, logged	0.077	0.170	0.034	
Sex of farm manager consistent across Afrint II and III with Afrint I household head, dummy	0.010	0.091	0.008	

	Model summary			
Model	r	r^2	Adjusted r^2	Std. error of the estimate
1	0.529a	0.280	0.211	0.40627

Notes: a. B is the symbol used for an unstandardized ordinary least squares regression coefficient.

two further variables that are influencing yields in Zambia: the trend in able workers (0.17) is significant at the 5 per cent level, and starting to use conservation agriculture (0.14) is significant at the 10 per cent level. In the same fashion as the Malawian model, the maize price variable and gender of farm manager show no significance.

Discussion of Results

For our discussion of the panel data results we draw on qualitative data collected in Malawi and Zambia in 2012 and again in Zambia in 2016. The production functions identified three independent variables as having a significant relationship with maize yields in both Malawi and Zambia: a positive relationship between fertilizer application and yields, and a relationship between distance to market/village centre and yields which is positive in Malawi (households further away from density/markets have greater yields) and negative in Zambia (households closest to density/markets have the greatest yields).

Despite the criticism that fertilizer subsidies have fostered, our data show that they have had an overall positive impact on maize yields, appearing to be the least-worst option for governments to support food security during an era of fluctuating global and national maize prices. In interviews, smallholders in both countries repeatedly and incisively return to describing households' fertilizer strategies. In Malawi poor soil fertility makes it imperative to apply appropriate amounts to achieve decent harvests. Smallholders rely almost exclusively on FISP, but when possible they buy additional unsubsidized fertilizers. Existing critique that the programmes create dependency and undercut the private market is confirmed. Village governments are in charge of distribution and as a rule they compel eligible households to share their bags of fertilizers, frequently causing respondents to complain that the amounts received are not sufficient. While FISP in Malawi is targeted to cater for the poorer households, the distribution principles for FISP in Zambia discriminate against those with less than 0.5 ha. Andersson Djurfeldt and Hillbom (2016) show how uneven access to subsidized fertilizers is part of an ongoing polarization process among Afrint households in Zambia. Smallholders depend on their memberships in cooperative unions for access; the more unions you join the more fertilizers you can access. While some cooperatives are active and members share multiple activities, others only exist for the distribution of FISP. The dependence on FISP and the programme's reoccurring late and unreliable distribution of fertilizers is a common complaint from respondents.

At first glance, the result from Malawi that households further away from markets and village centres have greater maize yields may not make intuitive sense. Taking into account the substantial subsistence production, variety of marketing opportunities, and alternative high-value crops, the result can, however, be explained. First, there are regional variations in the extent to which smallholders engage in the marketing of maize, many only growing maize for household consumption. Second, producers do not necessarily take their maize to the market themselves—they often prefer to sell to vendors at the farm gate thereby saving paying for transportation, costs that increase with distance. Furthermore, closeness to urban areas allows smallholders in densely populated Malawi to profit from the deepening of agricultural markets as they switch to high-value and more perishable crops such as vegetables, or provide for a growing 'processing industry', e.g. in the form of Irish potatoes or soya beans. Turning to the Zambian data, a different dynamic appears to be at work. While the FRA in many ways dominates the maize market and provides buying stations throughout the country, the parastatal's inability to pay farmers on time entices many to switch to regionally and locally established grain companies, cooperatives, and private buyers. The closer smallholders are situated to markets the more likely they are to profit from increasing competition as well as keeping transportation costs down. Moreover, the much lower population density in Zambia means thinner local markets and limited opportunities for smallholders to switch into high-value food crops as is the case in Malawi.

Further, we have significant results that are country specific. The strong negative relationship between maize production and size of maize area in Malawi resonates with intensification and increasing land scarcity emphasized in key informant interviews. Age proved to have a significant negative relationship with maize yields in Malawi. This is an interesting observation as one often hears arguments about youth lacking interest in farming. Here we see that it is the younger households with stronger labour who are increasing productivity to a greater extent than their older peers. In Zambia we obtain a different result in regard to access to labour resources. Here increasing maize production is based on extensification processes, with more land being dedicated to maize farming, but the production function shows that it is not only land which plays a role in this growth process. Access to labour in the form of more able workers, i.e. more family members able to participate in farm work, also turns out to be important.

Furthermore, we have the positive relationship between practising conservation farming and increasing maize yields in Zambia. This is corroborated in the qualitative data where conservation farming, specifically ripping, is claimed to provide multiple and substantial benefits. As one progressive farmer near Mkushi explains, practising ripping, or deep tilling of only the row where the crop will be planted, means that he no longer needs to plough and weed the whole field. Thereby he can farm a larger area compared to before without adding labour resources. Further, he can now plan his farming activities better, which reduces fluctuations in labour demand. Conservation farming practices have also increased his yields significantly. These claims are confirmed by the local agricultural extension officer.[3]

Before ending the discussion we need to comment on two insignificant results—prices and gender. We set up this study partly in the context of the recent commodity boom, and showed that national prices for white maize have experienced a rise (and fall) during the period of investigation in both countries. Subsequently, we hypothesized that rational smallholders would react by increasing their maize production, but this turned out not to be the case. What we can take from this result is that we should be careful when anticipating that smallholders living in remote areas in landlocked African countries will react to price incentives represented by global and national price levels. Finally, while households with female farm managers could not keep up with their male counterparts in terms of maize production they are sharing in the productivity increase, and their farming strategies are influenced by the same factors. This is a positive, and perhaps somewhat unexpected, result.

Concluding Remarks

In conclusion, a few results in our study are worth revisiting as they potentially carry policy implications. First, regarding existing agricultural policies it is clear that fertilizer subsidies have led to improved maize production. This conclusion does not contradict the critique that has been raised against fertilizer subsidy programmes, but the way forward would be to improve on the administration of existing programmes rather than to abandon them. Further, considering the success of conservation farming in both labour-saving farming methods and yields in Zambia, there may well be some scope for a systematic appraisal of conservation farming in Malawi. Third, while market prices offered by the FRA in Zambia have been beneficial for large-scale and emergent farmers, and contributed significantly to the growth in maize production, there are further market-enhancing reforms that remain to be

[3] This partly contradicts the claims of the labour intensity of conservation farming also in Zambia as pointed to in Chapter 3. The explanation appears to lie in the type of conservation farming being practiced, where the construction of basins is more labour intensive than ripping, especially if the latter is not done by hand.

addressed. Development of physical infrastructure and improved spread of price information would greatly assist smallholders by lowering transaction costs for market participation. It would make farmers less dependent on middlemen and more directly involved in markets, thereby more likely to profit from any future price hikes. Finally, considering their lower asset holdings, the ability of female farm managers in both countries, and younger households in Malawi, to share in broad productivity increases demonstrates how maize and food security policy should not marginalize such households.

9

Spatial and Gendered Linkages Between Non-Farm Diversification and Farm Productivity in Ghana

Fred Mawunyo Dzanku and Daniel B. Sarpong

Introduction

Increasing smallholder farm productivity in sub-Saharan Africa in general, and Ghana in particular, has been at the core of many agriculture policy interventions and the subject of several pieces of scientific research. Proponents of an agriculture-led rural poverty reduction strategy argue that increasing smallholder productivity is necessary for rural poverty reduction (World Bank 2007). Yet, the empirical literature suggests that rural households in developing countries straddle farm and non-farm employment (Davis et al. 2010, Dzanku 2015). A relevant research issue, then, is whether there exist household-level complementarities between the two livelihood strategies. This chapter, therefore, seeks to address this research issue by examining the relationship between rural non-farm diversification and productivity in agriculture. Importantly, we also seek to answer the question as to whether this relationship differs by the agro-productive potential and gender of non-farm income (NFI) earners within the household.

The relationship between rural non-farm diversification and farm productivity is ambiguous *ex ante*: on the one hand, the income generated from non-farm labour market participation could be used to fund the adoption of new technologies that improve yields, as well as mitigate the risk associated with adoption, particularly under rain-fed conditions. On the other hand, if household labour allocations towards rural non-farm employment (RNFE) draws labour away from the farm when the marginal physical product of on-farm labour is positive, and where hired labour markets are prohibitive, then farm output could decline. Depending on the magnitudes of these two effects, for example, the effect of non-farm diversification on farm productivity could be positive or negative, and this has to be determined empirically.

This study is not the first to examine the relationship between non-farm diversification and farm outcomes in general (see, for example, Evans and Ngau 1991, Ellis and Freeman 2004, Pfeiffer et al. 2009, Mathenge et al. 2015). Our contribution to the existing body of knowledge on farm–non-farm linkages lies in the use of longitudinal data, which is rare in the context of Ghana in particular. Probably more important, the analysis incorporates a gender dimension in a way that the extant literature does not. This is important because diversification into non-farm employment is not gender neutral; on the contrary, opportunities for non-farm diversification and the returns thereof differ by gender (see, for example, Corral and Reardon 2001, Elbers and Lanjouw 2001, Rijkers and Costa 2012, Ackah 2013). Naturally, therefore, an assessment of the linkages between non-farm diversification and farm productivity needs to account for the gendered nature of non-farm participation.

The gender analysis of the link between non-farm diversification and farm productivity is achieved through intra-household analysis. That is, we examine how, if at all, the effect of non-farm earnings on farm productivity differs by the gender of the income earner within the household. Thus, we hypothesize that the gender of the household member diversifying into RNFE matters for the effect on household agricultural productivity. Why might this be the case? Reardon et al. (1994) note that, among other factors, the investment of non-farm earnings into farming depends on who controls the NFI within the household; and our assumption is that who earns the income could be an important predictor of who controls it.

The rest of the chapter is organized as follows. The next section provides an overview of the country and local context. The data, theoretical framework, and the empirical models appear in the section after this. The results of the analyses and conclusions are in the two following sections respectively.

The Study Context

This section contextualizes the study area, first at the country level, and then at the regional and village levels. At the country level we focus on broad contextual issues during the years between 2000 and 2015 because the first round of the Afrint surveys utilized in the analysis was collected in 2002.

Country Context

At the beginning of the Afrint studies (in 2002, hereafter Afrint I), Ghana was designated a low-income country, according to the World Bank classification;

by the time of the most recent survey (2013, hereafter Afrint III) it had become a lower middle income (LMI) country. The LMI country status, which came about after the rebasing of GDP by the Ghana Statistical Service in November 2010, indicated that Ghana achieved this status in 2007 (i.e. around the time of the second round of the Afrint surveys, hereafter Afrint II).

Real GDP (RGDP) per capita figures during the Afrint II period represent a 19 per cent increase over Afrint I levels, and the levels attained during the period of Afrint III represent a 27 per cent rise over that of Afrint II. So, between the two points (i.e. Afrint I and III), RGDP per capita rose by about 52 per cent—from US$999 to US$1,521 (World Bank 2016). Since the year 2001, overall annual GDP growth rates have been 4 per cent and above, although growth has recently been sluggish (Figure 9.1). Ghana has also achieved significant gains in poverty reduction, from the initial Afrint I level of around 44 per cent (in 1999) to 32 per cent in 2006, and then to 24 per cent in 2013.

As one might expect, as in most other African countries south of the Sahara, agriculture's contribution to GDP in Ghana has been falling consistently, from around 40 per cent at the beginning of the 2000s to an average of about 22 per cent over the last five years (2011–15). In the context of the present study, non-farm activity and employment has thus increased at the national level. What is worrying is that although the economically active population employed in agriculture has been declining, as expected, the decline is slower compared with the decline in agriculture's GDP share, meaning that productivity has not increased as fast as one would expect. Agriculture remains the single largest

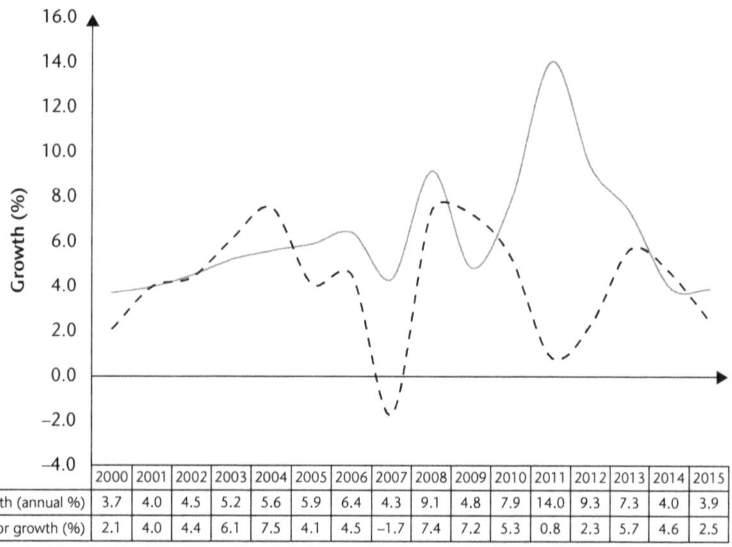

	2000	2001	2002	2003	2004	2005	2006	2007	2008	2009	2010	2011	2012	2013	2014	2015
——— GDP growth (annual %)	3.7	4.0	4.5	5.2	5.6	5.9	6.4	4.3	9.1	4.8	7.9	14.0	9.3	7.3	4.0	3.9
- - - - Agric. sector growth (%)	2.1	4.0	4.4	6.1	7.5	4.1	4.5	−1.7	7.4	7.2	5.3	0.8	2.3	5.7	4.6	2.5

Figure 9.1 Ghana's overall and agricultural GDP growth, 2000–15.

employer of labour, currently absorbing approximately 42 per cent of the labour force, down from 50 per cent at the beginning of the twenty-first century (Dzanku and Udry 2017). Indeed, as Figure 9.1 shows, average agricultural GDP growth has lagged behind overall growth over the period 2000–15 by about 2.1 percentage points.

An example of low agricultural productivity can be seen with cereals. The figures reported by the FAO and Ghana's Ministry of Food and Agriculture show cereal yields of about 1,268 kg/ha, 1,458 kg/ha, and 1,681 kg/ha during Afrint I, II, and III, respectively. These figures are well below LMI country standards by a factor of two, on average, while productivity-enhancing input use (e.g. chemical fertilizers) in Ghana also lags that of LMI peers. For example, fertilizer use (kg/ha of arable land) in Ghana is currently estimated at about 36, compared with 118 for the average LMI country. Even then, current levels represent a massive increase over Afrint I levels (less than 4 kg/ha).

On the policy front, the Ghana Poverty Reduction Strategies (GPRS I and II) contained government policy for the agriculture sector at the beginning of the Afrint studies, with the main focus being agricultural modernization and agro-industrialization. Details of the plans were contained in the Food and Agriculture Sector Development Policy documents (FASDEP I and II), and later, the Medium-Term Agriculture Sector Investment Plan which was designed for implementing the policies contained in the FASDEPs until 2015. A summary and analysis of these policies can be found in Dzanku and Aidam (2013). The general story is that although these policies appear good on paper, implementation challenges led to most of the sector growth targets being missed. There is thus very little evidence that agricultural modernization has begun to any large extent. One sign of this is that still less than 1 per cent of cropped land is under irrigation.

Study Sites

The study is located in eight villages spread across four districts in two of Ghana's administrative regions: Eastern Region (ER) and Upper East Region (UER). Importantly, the two regions are located in distinct agro-ecological zones. The UER villages are all in the Guinea Savannah Zone (GSZ); two of the ER villages are in the Semi-Deciduous Forest Zone (DFZ) and the other two in the Forest Zone (FZ). This clearly has implications for agricultural production potential and crop choice (or the lack thereof). This is more so because Ghana's agriculture is almost entirely rain-fed. In areas covering our study villages, annual rainfall amount and distribution varies in ascending order from the GSZ (800–1,200 mm) to the DFZ (900–1,500 mm) to the FZ (1,500–2,000 mm). Another important difference between the study sites is that whereas the ER villages have two rainy seasons in a year (major and minor based on duration), the UER villages have only one.

Sorghum, millet, rice, maize, and groundnut are the main crops grown in the UER villages while cassava, maize, plantain, and vegetables are important for the ER villages. In addition, the FZ villages grow tree crops such as cocoa and oil palm. Livestock production is much more important and widespread in the UER villages than in the ER. For example, in the nationally representative dataset 76 per cent of households in the UER owned livestock in 2013 compared with 45 per cent of households in the ER (84 per cent and 64 per cent, respectively, when considering rural households only).

The seasonal differences could have ramifications for non-farm participation given that farm and non-farm work could be sequenced in the UER while they may largely be concurrent in the ER. Among other factors, non-farm activities are known to be related to population density or urbanization and proximity to urban areas (Renkow 2007, Deichmann et al. 2008, Haggblade et al. 2010). The villages being studied also have varying population sizes and proximity to urban areas, all of which are important for participation in the rural non-farm sector. Two of the villages in the ER are essentially suburbs of their respective district capitals (Odumase-Krobo and Begoro); the other two are more removed from the capital. Two of the UER villages are within a 25 km radius of the regional capital, Bolgatanga; of the other two, one is on the Bolgatanga–Navrongo highway while the other is about 15 km from Navrongo, the capital of the Kassena Nankana Municipality. An important feature of this village is that it lies within the catchment area of the Tono Irrigation Project.

Methods

This section provides a description of the data employed for the analysis, motivates the empirical model by presenting a simple theoretical framework using the agricultural household model, and then outlines the empirical model and identification strategies. We begin with the data description.

Data

This study utilizes the Afrint panel datasets described in detail in Chapter 1. Details of the Ghana sample and distribution can be found in Dzanku and Sarpong (2014). The initial Afrint sample covered 416 households (in 2002) of which 357 were successfully followed in 2008, but a refreshment sample was also drawn in addition and so a total of 568 households were interviewed during Afrint II. A third round of the surveys took place in 2013 and involved

539 households, 87 per cent of whom were Afrint II households. Due to the availability data on our main variables of interest, our analysis uses the last two waves of the panel dataset (i.e. Afrint II and III), and the Afrint III cross-section.

Theoretical Framework

We motivate the empirical application as follows. We assume that households seek to maximize utility, U, defined over consumption of both own-produced and purchased goods (C) and leisure (l), given household-specific characteristics (Θ_{hh}). The household also supplies labour, L^n, to non-farm work which received wages, w, as well as receives exogenous non-farm non-labour income, V. In order to produce farm output, Q, the farm household uses land input, A, and purchased inputs, X, including hired labour, L^h. Production is also influenced by public goods and services, G (e.g. extension services), as well as agro-ecological characteristics, Z, which affect both the type of crop the household can cultivate and maximum attainable yields. Maximizing the Lagrangian associated with the household's utility maximization problem with respect to C and l and taking the ratio of the two expressions yields

$$w^* = \frac{(\partial U/\partial l)}{(\partial U/\partial C)} \geq w, L^n \geq 0, \quad \left(\frac{(\partial U/\partial l)}{(\partial U/\partial C)} - w\right) L^n = 0. \tag{9.1}$$

This expression is the marginal rate of substitution between consumption and leisure, w^*, and is an important first-order condition which states that each household equates their marginal rate of substitution between consumption and leisure (or shadow wage) either to their market wage or to their marginal product in farm work. Thus, w^* is endogenously determined due to the non-separability of production and consumption decisions. However, if L^n is observed, then w^* can be equated to the household's non-farm wage rate, w. This simple framework motivates household labour allocations between farm and non-farm livelihood activities.

Since some households may not supply any labour to non-farm activities, w may not be observed for all households. Because the aim here is to examine the effect of non-farm activity on productivity in agriculture rather than estimate directly household labour supply decisions, the study follows Sakurai and Reardon (1997) in introducing an NFI variable (*nfmy*) directly into the production function as:

$$Q = f\left(L^f, L^h, X, A, nfmy | \Theta_{hh}, Z, G\right). \tag{9.2}$$

In this context, non-farm diversification is considered an indirect household labour supply decision, and as shown intuitively and from the empirical literature, the non-farm diversification effect cannot be signed *a priori* because it could have either positive or negative effects on farm productivity in a similar way as crop diversification may have (Coelli and Fleming 2004).

Empirical Model and Measures

For the measure of farm productivity we focus on the total value of farm output per capita because we believe this is more directly related to household resource allocations between the farm and non-farm sector in the rural economy. Besides, it is mainly increases in labour productivity that could be expected to release labour for rural non-farm income-generating activities (Valdés and Foster 2007). In constructing the outcome variable we aggregate all crop output as determined through local markets. We then use the implied village average price to impute the value of own crops used for other purposes other than sale (i.e. home consumption, payment for hired labour, transfers to relatives, seed, animal feed, brewing, gifts, etc.). A limitation with our measure is that because our questionnaire did not capture detailed production data for some important food crops in the study areas, such as cassava, plantain, and groundnuts, only those marketed could be captured in the output measure. The general form of the productivity equation is specified as

$$\Delta y_i = \alpha_0 + \delta \Delta nfmy_i + \Delta X_i \beta' + W_i \eta' + u_i, \tag{9.3}$$

where Δ represents change in the respective time-varying variables between Afrint II and Afrint III for household *i*, *nfmy* is non-farm earnings; *X* contains a vector of time-varying covariates of farm productivity described in equation 9.2, *W* is the vector of time-invariant control variables, and u_i is the idiosyncratic error which captures random production shocks that we did not measure, for example, weather and other risk factors; δ, β, and η are all parameters to be estimated, with our main interest being δ. For the estimate that uses gender-specific non-farm earnings, equation 9.3 is modified as

$$y_i = \alpha_0 + \delta_1 nfmy\_w_i + \delta_2 nfmy\_m_i + X_i \beta' + u_i, \tag{9.4}$$

since the gender-specific non-farm earning variables (*nfmy_w* and *nfmy_m*, for women and men, respectively) were measured during Afrint III only. Thus, for equation 9.4 the vector *X* contains all the controls.

In our panel data estimates, inputs and all other time-varying covariates are *ex ante* allowed to be correlated with household unobservables, meaning the fixed effects (FE) estimator is preferred. The specification in equation 9.3 is essentially equivalent to the FE estimator for $T = 2$ (Wooldridge 2002: 284), as is the case for the current application.

Finally, we devise an identification strategy for dealing with the possibility that non-farm earning is endogenous given that non-farm participation is not randomly assigned. Since non-farm earning enters the models as semi-continuous (or censored), we follow the control function approach of Vella (1993). This procedure uses the generalized residual obtained from a Tobit regression of *nfmy** on all exogenous variables (including the full set of instruments) as an additional regressor in equations 9.3 and 9.4. The null hypothesis that this additional regressor is equal to zero is the relevant exogeneity test. Due to the two-step nature of the procedure we bootstrap the standard errors with 1,000 draws. An alternative approach to dealing with potential endogeneity is full information joint estimation, but both approaches yield similar results in our case.

We use two excluded instruments that enter the first-step Tobit regression: connection to electricity supply and the presence of non-farm productive assets (mainly sewing machines). The justification for these instruments is as follows. The literature suggests that infrastructural development enhances non-farm diversification (Corral and Reardon 2001, Reardon et al. 2007, Gibson and Olivia 2010, Losch et al. 2013). One such key infrastructure is electricity, the absence of which limits non-farm activity (Rijkers and Costa 2012, Mathenge et al. 2015). Similarly, the sewing machine is a key non-farm asset in rural Ghana. While these variables should be sufficiently correlated with non-farm diversification, we do not expect that they independently explain farm production and productivity except through non-farm earnings.

Results

Descriptive Analytical Results

We begin the discussion of the descriptive analytical results focusing on the dependent variable and the key explanatory variable of interest (Table 9.1). We compare these variables across gender of farm manager and between regions.[1] The literature documents gender gaps in farm productivity in favour of males (Goldstein and Udry 2008, FAO 2011, Kilic et al. 2015, Slavchevska 2015) with gaps in the region of 4 per cent to 40 per cent. The descriptive evidence from our data shows mixed results. The productivity gap in the pooled data is in favour of female farm managers (FFMs) (1.7 per cent) but not statistically different from zero. This result is due to two opposing

[1] We use sex of farm manager in all the analysis related to gender. FFM and MFM denotes female and male farm manager, respectively, whereas FMF and MMF are as defined earlier. In the balanced pooled Afrint II and III sample used for the present analysis, the sex of the farm operator/manager is the same as that of the household head in 85 per cent of cases.

Table 9.1. Farm labour productivity and non-farm income across genders and region over time

Survey round	Household categories	Value of farm output per capita (US$)	Non-farm labour market participation rate	Real non-farm earnings (US$)	Real conditional non-farm earnings (US$)	Intra-HH non-farm earnings-female US$)	Intra-HH non-farm earnings-male (US$)
Afrint II and III	All HHs	138	54%	232	431	n.a.	n.a.
	FMF HHs	140	56%	181	326	n.a.	n.a.
	MMF HHs	138	53%	246	460	n.a.	n.a.
	Diff.	−2	−2%	64	134**	n.a.	n.a.
Afrint II	All HHs	120	42%	201	478	n.a.	n.a.
	FMF HHs	154	42%	166	400	n.a.	n.a.
	MMF HHs	112	42%	210	496	n.a	n.a.
	Diff.	−42*	1%	43	97	n.a.	n.a.
Afrint III	All HHs	157	66%	263	402	n.a.	n.a.
	FMF HHs	128	69%	195	284	n.a.	n.a.
	MMF HHs	165	65%	282	436	n.a.	n.a.
	Diff.	37*	−4%	87*	151**	n.a.	n.a.
Afrint II and III	ER HHs	244	42%	279	659	n.a.	n.a.
	UER HHs	56	63%	196	311	n.a.	n.a.
	Diff.	188***	−21%***	83**	348***	n.a.	n.a.
Afrint II	ER HHs	235	30%	216	717	n.a.	n.a.
	UER HHs	30	52%	190	368	n.a.	n.a.
	Diff.	205***	−22%***	26	349***	n.a.	n.a.
Afrint III	All HHs					368	593
	ER HHs	253	55%	342	627	450	675
	UER HHs	81	74%	201	272	298	524
	Diff.	172***	−20%***	141***	355***	151***	152

Notes: a. Diff. denotes the gap between MMF and FMF households (i.e. MMF HHs − FMF HHs) or the gap between ER and UER households (i.e. ER HHs − UER HHs).
b. n.a. means not available.
c. *, **, and *** denote statistical significance at the 10 per cent, 5 per cent, and 1 per cent levels.
d. HH stands for household.

outcomes: a significant gender gap in favour of FFMs in Afrint II (about 27 per cent), and a significant gap in favour of male farm managers (MFMs) in Afrint III (about 29 per cent).[2]

We expected significant regional differences in productivity owing to superior agro-ecological conditions under rain-fed production systems in the ER compared with the UER. This is what we observe. Productivity in the pooled sample is higher in the ER than in the UER by approximately a factor of four (or a gap of about 339 per cent). We note, however, that the productivity gap narrowed markedly over time, from a factor of approximately eight during Afrint II to a factor of about three during Afrint III.

[2] Unless otherwise specified, we use the word significant to connote statistical significance at conventional levels.

We now turn to the main explanatory variables of interest: non-farm diversification and earnings. We define non-farm diversification as participation in rural non-farm labour markets, thus excluding non-labour income (mainly remittances and transfers). Table 9.1 shows that non-farm diversification increased from 42 per cent participation rate in Afrint II to 66 per cent in Afrint III, the time effect is significant at the 1 per cent level. Over the same period, real non-farm earnings grew by approximately 31 per cent (about 5 per cent a year), but this growth in earnings is not statistically significant. Accelerated growth of the services sector in the economy of Ghana as a whole was observed between 2008 and 2013.

Based on the existing literature (e.g. Canagarajah et al. 2001, Ackah 2013), one would expect non-farm diversification to be higher among women than men. Overall, we find no significant gender gap in both non-farm labour market participation and earnings in our sample. However, for Afrint III we observe a gap of approximately 45 per cent in earnings in favour of MFM households. Looking at non-farm earnings conditional on participation, we see a significant gender gap in favour of MFM households in the pooled sample as well as for Afrint III. This could be suggestive of females, on average, being engaged in less remunerative non-farm work than males.

Turning to regional differences in non-farm activity, we observed a significant gap in participation, with rates higher in the UER than in the ER by about 21 per cent in the pooled sample. Similar gaps in participation rate is observed for Afrint II and Afrint III. This is not surprising given the longer agricultural off-season in the UER, for which some household members migrate in search of seasonal non-farm employment. Although participation rates are higher in the UER, average earnings are significantly lower, with a significant average earnings gap of about 42 per cent in favour of ER households in the pooled sample. The pooled sample gap was driven mainly by a large gap of approximately 70 per cent in Afrint III. The average gap between the two regions is even wider when we consider average earnings conditional on participation.

The last two columns of Table 9.1 consider intra-household differences in non-farm earnings in the full sample and across regions. Overall, average earnings are significantly higher for males than females, with an average gap of about US$226 (or approximately 62 per cent). The intra-household gender gap persists in both regions. When we consider differences in female non-farm earnings between the two regions we observe that women in the ER earn significantly higher incomes. On the other hand, analysing male intra-household income differences across the regions shows that although there exists an average gap of approximately 29 per cent in favour of men in the ER this difference lacks statistical significance at conventional levels.

Next, Figure 9.2 plots the correlation between farm labour productivity and non-farm diversification probabilities. A general positive association is

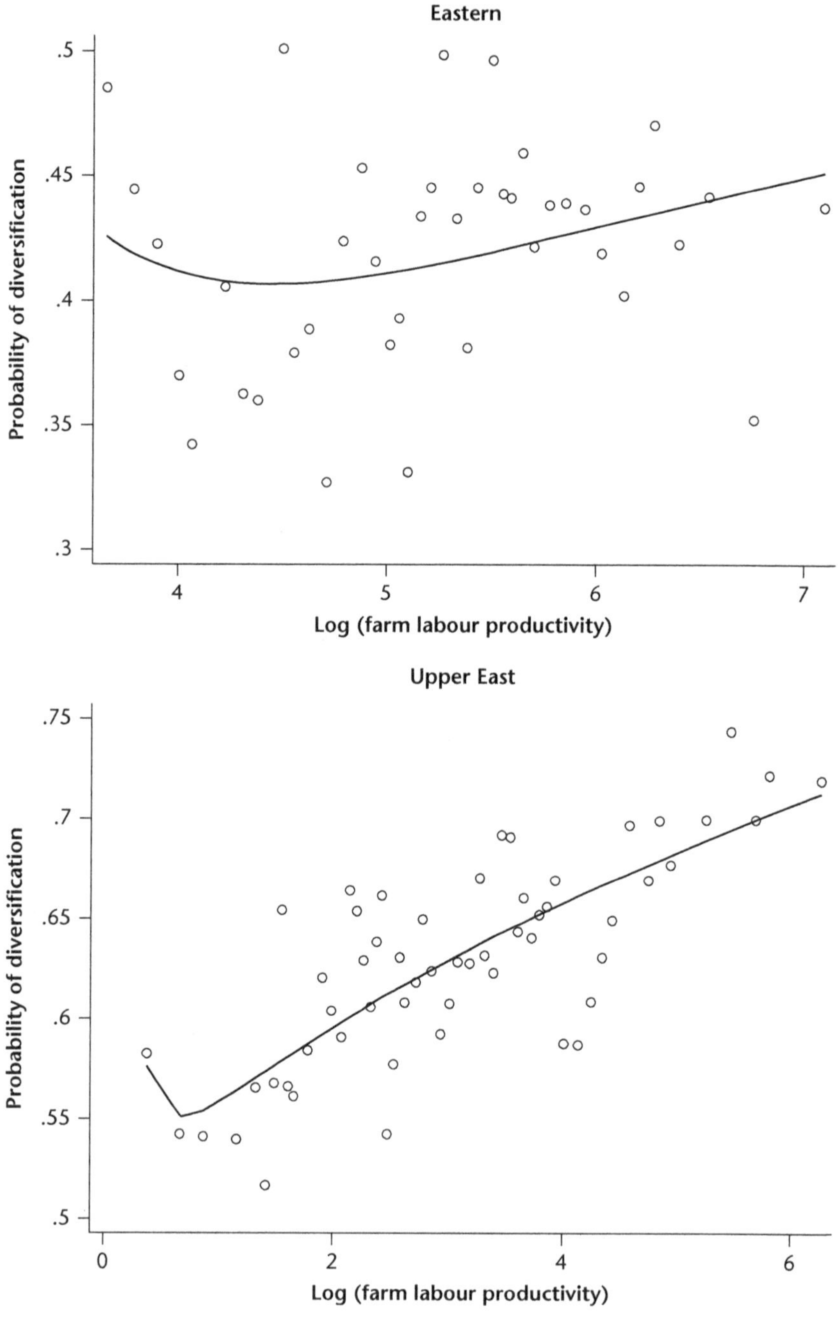

Figure 9.2 Non-farm diversification and farm labour productivity, by region.

observed for both regions, but this shows up much more strongly in the UER (lower panel of Figure 9.2) than in the ER (upper panel of Figure 9.2). This could be related to the different agricultural seasonality patterns in the two regions. In the next subsection, we control for other possible confounders, including unobserved household heterogeneity and the possibility of endogenous diversification behaviour using regression analysis.

Before then, Table 9.2 presents summary statistics of variables used in the regressions for the pooled sample (Afrint II and III), and for each round of the surveys (Afrint II and III). We also provide in the last column statistical tests of

Table 9.2. Descriptive statistics of variables used in the regression analyses

Variable	Afrint II and III	Afrint II	Afrint III	Afrint III – Afrint II
Value of farm output per capita (2010 US$)	138.39	119.90	156.89	36.99***
Non-farm earnings (2010 US$)	232.42	201.35	263.49	62.14*
Women's non-farm earnings (US$)			386.02	n.a
Men's non-farm earnings (US$)			535.57	n.a
Cultivated area (ha)	2.18	1.79	2.56	0.77***
Household labour input (persons)	3.92	4.14	3.69	−0.45***
Hired in-farm labour (yes)	66%	63%	69%	7%**
Used improved seeds (yes)	44%	60%	29%	−31%***
Used chemical fertilizer (yes)	57%	53%	60%	8%**
Improved seed-fertilizer use				
Uses neither improved seed nor chemical fertilizer (yes)	29%	25%	32%	7%**
Uses improved seed but not chemical fertilizer (yes)	15%	22%	8%	−14%***
Use chemical fertilizer but not improved seed (yes)	27%	15%	39%	24%***
Used both improved seed and chemical fertilizer (yes)	30%	38%	21%	−17%***
Use of irrigation water	35%	38%	32%	−6%**
Used pesticides (yes)	33%	34%	33%	−1%
Received agricultural extension advice (yes)	51%	53%	49%	−3%
Belongs to farmer association (yes)	19%	22%	17%	−5%**
Livestock wealth (tropical livestock unit)	1.88	1.92	1.84	−0.08
Perception of improvement in output market prices (yes)	26%	22%	31%	9%***
Has secure land rights (yes)	77%	74%	80%	5%**
Age of farm manager (years)	50	49	52	2.79***
Female farm manager (yes)	20%	20%	21%	2%
Farm manager's years of schooling	4.8	5.0	4.7	−0.33
Instruments				
Connected to electricity (yes)	22%	14%	30%	16%***
Presence of non-farm assets (yes)	26%	28%	24%	−4%
Ecological zones				
Guinean Savana Zone (yes)	56%	56%	56%	
Deciduous Forest Zone (yes)	23%	23%	23%	
Forest Zone (yes)	21%	21%	21%	
Regions				
Eastern Region (yes)	44%	44%	44%	
Upper East Region (yes)	56%	56%	56%	
N	982	491	491	

Note: a. *, **, and *** denote statistical significance at the 10 per cent, 5 per cent, and 1 per cent levels.

differences over the two waves (i.e. Afrint III – Afrint II). We observe that mean farm labour productivity increased significantly over the two waves, averaging a growth rate of about 5 per cent per year over the six-year period.

The farm productivity models control for traditional input variables, namely land, labour, chemical fertilizer, irrigation, pesticides, type of planting material, agricultural extension and farmer association membership, and household head characteristics. Location FEs are included to control for local climatic factors and other seasonal and spatial shocks and unobservables. Table 9.2 shows that while there was a decrease in some of the factors of production over time others experienced increases. The following experienced a significant decline over the period: household labour input (10.9 per cent), the proportion of farmers using improved planting materials (30.9 per cent), the proportion of farmers using both improved seeds and chemical fertilizers (16.7 per cent), the proportion of farmers using irrigation water (6.3 per cent), and the proportion of farmers belonging to farmer associations (5.1 per cent). On the other hand, the following increased significantly between Afrint II and Afrint III: the proportion of farmers using chemical fertilizers (7.5 per cent), total cultivated area (43.1 per cent), and the proportion of households hiring farm labour (6.7 per cent). The proportion of farmers using pesticides, and the proportion of farmers who had contacts with agricultural extension services, experienced no significant change over the period.

Regression Results and Discussion

CHOICE OF ESTIMATOR AND ROBUSTNESS ISSUES

In the full sample estimation of the farm labour productivity equation the robust test of overidentifying restrictions returned a Sargan–Hansen statistic of 116.593 (with 16 degrees of freedom), which clearly rejects the null hypothesis that the coefficients from FE and random effects (RE) are not systematically different (p-value = 0.000). The first-differencing procedure applied (equation 9.3) with the modification to accommodate time-invariant covariates is thus justified.

Although we believe our instruments are defensible based on the received literature, we carried out a statistical test to verify whether they are statistically relevant and valid. We use a linear specification where we estimate equation 9.3 by two-stage least squares. The resulting underidentification test—a test of the null hypothesis that the instruments are uncorrelated with non-farm earnings—easily rejects the null with zero p-values to three decimal places. The joint F-test of the strength of the excluded instruments also shows that they are strongly correlated with the endogenous regressor. The p-value from the Hansen J statistic (0.518) suggests that the overidentifying restrictions are valid (i.e. the instruments are uncorrelated with the error term in the 'structural equation'). We now turn to the discussion of our regression results.

NON-FARM DIVERSIFICATION AND FARM PRODUCTIVITY: REGION-SPECIFIC PERSPECTIVES

Economic theory suggests that labour movements from agriculture to non-agriculture could have a positive effect on agriculture in the presence of an underdeveloped agricultural sector with surplus labour and declining marginal physical product. If there are no labour surpluses, however, the effect could be hard to sign. Tables 9.3 and 9.4 report the coefficient estimates from equations 9.3 and 9.4 respectively. In both cases, we report full sample results as well as regionally disaggregated estimates. The justification for estimating separate regressions for each region is borne out of the large differences in farm production conditions as well as non-farm labour market differences evidenced in Table 9.1. More formally, we carried out a likelihood ratio test for the null hypothesis that all coefficients of the farm productivity models do not vary between the ER and UER. The chi-squared statistic for this hypothesis with 17 degrees of freedom equals 65.56 (p-value = 0.000). The corresponding statistic for equation 9.4 is 80.26, with 20 degrees of freedom (p-value = 0.000).

In all the control function estimates (CFE) reported in Table 9.3, none of the coefficients on the correction factor, NFYGRES, is statistically different from zero, meaning that there is insufficient evidence from our data to reject the null hypothesis that non-farm earning is endogenous in the farm labour productivity equations. All coefficients on non-farm earnings are negative except in the CFE for the ER. Nonetheless, only the OLS estimates for the UER shows statistical significance at the 5 per cent level. We also estimated equation 9.3 using the value of farm output rather than output per capita as a dependent variable and the results (available on request[3]) are very similar to those reported in Table 9.3. The negative non-farm earnings effect observed from our data is similar to the finding of Pfeiffer et al. (2009), who used cross-sectional data on Mexico. This finding is suggestive of non-farm earning opportunities competing rather than complementing agricultural production, although the evidence is far from conclusive.

The overall lack of a significantly strong association between farm labour productivity and non-farm earnings may be suggestive of the availability of surplus household labour such that allocating time off-farm has no productivity-reducing effect on-farm. What is surprising, however, is the significant negative effect of non-farm earnings on farm labour productivity in the UER where one would have expected the relatively longer agricultural

[3] Please contact the lead author, via fdzanku@gmail.com or <http://freddzanku.com/>.

Table 9.3. Effect of non-farm income on farm labour productivity, Afrint II and III balanced panel

Variables	Full sample		Eastern Region		Upper East Region	
	OLS	CFE	OLS	CFE	OLS	CFE
Δ Non-farm earnings (log)	−0.024	−0.026	−0.007	0.007	−0.044**	−0.073
Δ Cultivated area	0.124***	0.124***	0.047	0.048	0.114**	0.120**
Δ Cultivated area squared	−0.017**	−0.017**			−0.027*	−0.027*
Δ Household labour input	−0.063***	−0.063***	−0.085***	−0.086***	−0.039*	−0.040*
Δ Hired in-farm labour	0.274***	0.275***	0.388**	0.379**	0.147	0.149
Improved seed-fertilizer use (ref. = none)						
Δ Improved seed only	0.320**	0.320**	0.364	0.366	0.099	0.098
Δ Chemical fertilizer only	0.411***	0.412***	0.320	0.316	0.072	0.080
Δ Improved seed plus fertilizer	0.152	0.152	0.368*	0.373*	−0.134	−0.130
Δ Used irrigation	0.077	0.077	−0.149	−0.148	0.557***	0.561***
Δ Used pesticides	0.269**	0.269**	0.069	0.061	0.506***	0.501***
Δ Received agricultural extension advice	0.042	0.041	−0.029	−0.025	−0.096	−0.107
Δ Member of farmer association	0.028	0.029	−0.084	−0.095	0.022	0.018
Δ Livestock wealth	−0.002	−0.002	−0.001	−0.004	−0.019	−0.018
Δ Improved output market prices	0.377***	0.378***	0.306**	0.302**	0.503**	0.506**
Δ Has secure land rights	−0.223**	−0.224**	−0.130	−0.122	−0.169	−0.183
Age of farm manager	−0.004	−0.004	−0.007	−0.006	−0.003	−0.003
Female farm manager	−0.189	−0.188	−0.200	−0.203	−0.079	−0.076
Farm manager's years of schooling	0.007	0.007	0.008	0.007	0.004	0.005
Ecological zone (ref. = Guinea Savanna)						
Deciduous forest zone	−0.685***	−0.685***				
Forest zone	−1.213***	−1.214***				
Deciduous forest zone vs forest zone			0.450**	0.447**		
Bolgatanga area vs Navrongo area					−0.903***	−0.900***
Constant	1.067***	1.073***	−0.012	−0.041	1.526***	1.563***
NFYGRES		0.004		−0.045		0.072
Observations	491	491	215	215	276	276
Adj. *R*-squared	0.262	0.262	0.213	0.213	0.280	0.280
F-stat	8.291	7.906	3.346	3.170	5.919	5.683
p-value	0.000	0.000	0.000	0.000	0.000	0.000

Notes: a. Inference is based on heteroscedasticity consistent (HC3) standard errors.
b. *, **, and *** denote statistical significance at the 10 per cent, 5 per cent, and 1 per cent levels.

Table 9.4. Effect of gendered intra-household non-farm earnings on farm labour productivity

Variables	Full sample		Eastern Region		Upper East Region	
	OLS	CFE	OLS	CFE	OLS	CFE
Women's non-farm earnings (log)	−0.029*	−0.023	−0.049**	−0.062	−0.013	−0.038
Men's non-farm earnings (log)	0.001	−0.097*	0.044**	0.032	−0.046**	−0.073**
Cultivated area	0.472***	0.442***	0.368***	0.363***	1.010***	0.971***
Cultivated area squared	−0.050***	−0.046***	−0.039***	−0.039***	−0.145***	−0.137***
Household labour input	−0.084***	−0.079***	−0.091***	−0.090***	−0.010	−0.001
Hired in-farm labour	0.249**	0.230*	0.366*	0.384*	0.208*	0.157
Used improved seed					−0.180	−0.177
Used chemical fertilizer					0.071	0.170
Improved seed/fertilizer use (ref. = none)						
Improved seed only	0.723***	0.839***	0.316	0.318		
Chemical fertilizer only	1.048***	1.126***	0.558**	0.568**		
Improved seed plus fertilizer	0.927***	0.973***	0.469*	0.458		
Used irrigation	0.405***	0.398***	−0.007	−0.011	0.675***	0.688***
Used pesticides	0.122	0.150	−0.200	−0.183	0.123	0.087
Received agricultural extension advice	0.091	0.086	0.178	0.184	−0.148	−0.113
Member of farmer association	−0.033	0.004	0.106	0.118	0.134	0.154
Livestock wealth	0.008	0.013	0.204**	0.217*	0.006	0.008
Improved output market prices	0.526***	0.532***	0.470***	0.477***	0.649***	0.782***
Has secure land rights	0.018	−0.001	0.029	0.020	0.144	0.049
Age of farm manager	−0.002	−0.006	−0.008	−0.008	0.000	−0.004
Female farm manager	−0.072	−0.241	−0.107	−0.102	0.120	0.049
Farm manager's years of schooling	−0.000	0.003	−0.002	−0.001	0.001	−0.001
Ecological zone (ref. = Guinea Savanna)						
Deciduous forest zone	1.013***	0.874***				
Forest zone	0.693***	0.477*				
Deciduous forest zone vs Forest zone			0.220	0.230		
Bolgatanga area vs Navrongo area					−1.469***	−1.451***
Intercept	2.602***	3.117***	4.239***	4.281***	2.897***	3.128***
NFYGRES_W		−0.002		0.019		−0.021
NFYGRES_M		0.232*		0.042		−0.148
Observations	536	536	246	246	290	289
Adj. *R*-squared	0.540	0.544	0.293	0.293	0.642	0.646
F-stat	37.15	34.07	5.210	4.753	30.75	28.71
p-value	0.000	0.000	0.000	0.000	0.000	0.000

Notes: a. Inference is based on heteroscedasticity consistent (HC3) standard errors.
b. *, **, and *** denote statistical significance at the 10 per cent, 5 per cent, and 1 per cent levels.

off-season to allow sequencing of farm and non-farm activities such that, at worse, one has no adverse effect on the other.

NON-FARM DIVERSIFICATION AND FARM PRODUCTIVITY: INTRA-HOUSEHOLD GENDER-SPECIFIC PERSPECTIVES

Table 9.4 provides estimates of the association between intra-household gendered-specific non-farm earnings and farm labour productivity. First, we note that the coefficient on the correction factor from the men's non-farm earnings equation is positive and statistically significant in the full sample estimates. This means that unobserved factors that increase men's non-farm earnings also increase household farm labour productivity. The opposite is true for women's non-farm earnings, although this is not statistically significant. Once regionally disaggregated, the correction factors return insignificant coefficient estimates, suggesting that one cannot reject the null hypothesis of exogeneity of gender-specific non-farm earnings.

Turning to the gender-specific coefficient estimates, we observe (for the CFE) from the full sample results that both women's and men's non-farm earnings are negatively associated with per capita household farm output, but only the men's earning coefficient is mildly significant—a 1 per cent increase in men's non-farm earnings is associated with a 0.1 per cent decrease in total value of farm output per capita, *ceteris paribus*. Looking at the region-specific estimates we see that the full sample result was dominated by what pertains in the UER. In the ER, men's non-farm earning is associated with increasing farm output per capita, indicating complementarity between farm output and men's non-farm earnings. The opposite is true for women's non-farm earnings and farm output in the ER. Both effects are, however, not economically strong given the small elasticities.

Explaining the differing effects of women's and men's non-farm earnings on farm production requires further research. However, we note that in the ER villages, commerce, the dominant non-farm earning activity, is considered the preferred employment option for women. In households where women are engaged in commerce and related non-farm work, farming is considered the main occupation of men, even if they are involved in some non-farm work themselves (mainly artisanal jobs). This orientation means that it is men's NFI that is most likely to be used for purchasing inputs that could enhance farm productivity while women's engagement in non-farm work could drain the household of farm labour, leading to a negative farm productivity effect. In the UER, where returns to farming are much lower owing to the precarious agro-ecological conditions in the absence of irrigation and soil amendments, non-farm work (including seasonal migration to southern Ghana) is considered an alternative livelihood activity rather than one that complements farming through the farm investment channel.

These cross-sectional results (Table 9.4) should be interpreted with some caution given that they could still suffer from endogeneity due to omitted variable bias arising from unobserved household-specific heterogeneity. But all cross-sectional estimates using observational data are prone to this problem.

OTHER FACTORS

Before closing this section, we briefly discuss other covariates in the estimated equations. First of all, we note that the models have good explanatory power for a two-period panel (Table 9.3) and cross-sectional data (Table 9.4). Also, most of the explanatory variables are precisely estimated and are signed as expected, particularly for the ER equations. In the interest of brevity, we comment on only selected variables from the region-specific panel data results reported in Tables 9.3: farm size, labour input, modern input use, irrigation, agricultural extension contact, membership of farmer associations, market opportunities, and gender.

A well-known empirical regularity in agricultural production economics is the inverse relationship between output per unit of land and farm size. Our estimates using land productivity as a dependent variable confirms this empirical regularity. When it comes to labour productivity, however, while the relationship is positive but not significant in the ER it is significantly non-monotonic in the UER (joint F-stat = 3.93, p-value = 0.021)—labour productivity is first increasing with farm size up to 2.2 ha and then begins to decline with farm size. We note that in the UER, approximately 40 per cent of farm households in our sample have farm sizes above this threshold. We observe a significant inverse relationship between farm productivity and labour input in the ER but not the UER: *ceteris paribus*, an additional household labour input is associated with a 9 per cent decrease in farm labour productivity in the ER. On the other hand, hired labour use is positively associated with labour productivity.

We note that the relationship between chemical fertilizers and improved seed on the one hand and farm productivity on the other is not significant in the UER. The use of irrigation is more important in such semi-arid conditions: farmers in the UER using irrigation water are estimated to achieve 64 per cent higher output per capita than farmers producing under only rain-fed conditions. We observe no significant irrigation use effect in the ER. In this region, however, farmers combining improved planting materials with chemical fertilizers are estimated to achieve 53 per cent higher yields than those using neither. But the null hypothesis that productivity does not differ between those combining the two inputs and those using either one or the other could not be rejected. The other input which shows up as a significant positive predictor of productivity is the use of pesticide, but this is true only in the

UER where it is estimated to raise productivity by approximately 61 per cent over and above that of farmers who use no pesticides.

A few studies have analysed the association between market participation or commercialization and farm productivity using various approaches (Govereh and Jayne 2003, Ochieng et al. 2016). This suggests synergies between the two. Owing to the endogeneity of commercialization or market participation in a productivity model we attempt gauging the association using farmers' perception of output market price movements over time. We find that markets do matter for farm productivity irrespective of region. The perception of improvement in output market prices tends to be significantly associated with increased productivity.

Finally, our results show no significant association between average farm productivity and farm manager characteristics such as gender, age, and education after controlling for inputs and location FEs.

Conclusions

We now return to the main objectives of the study and relate them to the empirical findings. This study set two main objectives: (1) to examine the relationship between non-farm diversification and farm labour productivity; and (2) to investigate whether the gender of the person who diversifies into RNFE within the household matters for the effect, if any, of non-farm diversification on farm productivity. The novelty with this study and its contribution lies mainly in providing longitudinal evidence from Ghana as well as the intra-household gender perspective which has not been previously studied to the best of our knowledge.

After allowing for household-specific heterogeneity to be correlated with diversification behaviour, we find no evidence that non-farm earning increases farm labour productivity in the full sample. However, we find evidence from the region-specific estimates (which also correspond to differing agro-ecological conditions) that increasing non-farm earnings reduces average farm productivity in the region with poorer agricultural productive potential. On the hypothesis that the gender of who within the household diversifies into non-farm employment matters for the effect that diversification might exert on farm productivity, we find that our data provide some support for this hypothesis but that the relationship differs by region. In the relatively richer region (both in terms of agro-ecological potential and infrastructure endowment), farm labour productivity is significantly decreasing with women's non-farm earnings but increasing with men's, although the magnitude of the effect suggests that the association is not of strong economic significance. In the poorer region with a unimodal rainfall pattern and a longer off-season during which labour migrates in search of non-farm jobs, only men's non-farm earnings exert a significant negative effect on household farm labour productivity. However, we caution that the cross-sectional relationships could well be due to any remaining endogeneity bias attributable to omitted household-specific heterogeneity. Collecting intra-household gender-specific data on diversification over time could help shed more light on this hypothesis.

10

Technology Use, Gender, and Impact of Non-Farm Income on Agricultural Investment: An Empirical Analysis of Maize Production in Two Regions of Kenya

Stephen K. Wambugu, Joseph T. Karugia, and Willis Oluoch-Kosura

Introduction and Problem Overview

Agriculture continues to be a fundamental instrument for sustainable development, poverty reduction, and enhanced food security in developing countries. Agricultural productivity levels in sub-Saharan Africa are far below those of other regions in the world, and in most countries are well below the level required to attain food security and poverty reduction goals. Nonetheless, the rate of agricultural productivity has been growing since the early 2000s in many African countries, including Kenya, although this is not a cause for complacency (Kibaara et al. 2009). Sustained and accelerated growth requires a sharp increase in productivity among smallholder farmers. The national Strategy to Revitalize Agriculture (Republic of Kenya 2004), Kenya Vision 2030 (Republic of Kenya 2007), Comprehensive Africa Agricultural Development Program, 2013 (NEPAD 2013), and the Alliance for Green Revolution in Africa (AGRA 2014) have underscored the importance of increasing agricultural productivity in the fight against poverty. In the past, agricultural production was largely a function of acreage expansion, but further growth in production will have to be driven by productivity growth given that the land frontier has almost been exhausted in most countries. Adoption and continued use of improved agricultural technologies can play a big role in enhancing productivity especially for food staples (Kibaara et al. 2009). Productivity improvement can also be achieved through investing non-farm income (NFI) in agriculture, particularly through purchasing improved farm inputs.

NFI has been used to boost agricultural productivity when it is used for agricultural investment generally and in purchasing agricultural inputs specifically (Haggblade et al. 1989, Hazell and Hojjati 1995). Conversely, farm income can also be invested in non-farm activities. The importance of NFI for the livelihood strategies of rural people has attracted much attention among development scholars, policymakers, and donors during the past decade. Although NFI on an aggregate level is important in the rural economies of sub-Saharan Africa, the distribution of such incomes is normally skewed in favour of the better-off members of the population (Babatunde and Qaim 2009). The bulk of studies on income diversification out of agriculture have focused on mechanisms that can lower entry barriers and increase participation of the poor in such income generation. In general terms, not much attention has been devoted to the question: 'How do non-farm activities affect farming?' This is a pertinent question given that the great majority of rural Africans are increasingly deriving their livelihoods from NFI (Haggblade et al. 2010). Specifically, the link between NFI diversification and agricultural technology adoption and use has not received enough empirical investigation.

Non-farm activities have become important components of livelihood strategies among rural households in most developing countries. Several studies (Ruben 2001, De Janvry and Sadoulet 2005, Haggblade et al. 2007, Jirström et al. 2012, Wambugu and Karugia 2014, Alobo Loison 2016) have reported a substantial and increasing share of NFI in total household income. Reasons for this observed income diversification include declining farm incomes and the desire to insure against agricultural production and market risks (Reardon 1997, Kijima et al. 2006, Matsumoto et al. 2006). It has been observed that when farming becomes less profitable and more risky as a result of various factors—including climate change and variability, population growth, and crop and market failure—households are pushed into non-farm activities, leading to 'distress-push' diversification (Babatunde and Qaim 2009). In other cases, however, households are pulled into the non-farm sector, especially when returns to non-farm employment are higher or less risky than in agriculture, resulting in 'demand-pull' diversification (Babatunde and Qaim 2009).

Often it is unclear whether and how non-farm activities can contribute to equitable development and to bridging the gender gap. The gendered aspects of NFI diversification, and technology adoption and use (especially for the main staples), have not received adequate empirical investigation, an aspect that this chapter sets out to address. The chapter uses panel datasets collected in two regions (counties) of Kenya to analyse gendered patterns of technology use in relation to maize productivity and total household production. The chapter also assesses the impact of NFI on agricultural investments, zeroing in on fertilizer use (a key agricultural input) for maize production in Kenya. These aspects are disaggregated by gender of the farm manager to capture the gender dimensions. Maize has been chosen because it is the major staple food crop in Kenya.

This chapter addresses a number of research questions. The first research question is: Comparing female-managed farms (FMFs) and male-managed farms (MMFs) which ones achieve higher maize yields? The second question is: How technology adoption and use explain any observed differences in maize production? The third research question relates to how NFI is invested in agriculture. While the findings presented in this chapter are specific to the particular setting in two regions in Kenya, they may also contribute to a better general understanding of the underlying issues and linkages.

Agricultural Production and Gender

Agricultural production is the preoccupation of rural households involving men and women. Men and women are also involved in a wide range of farm and non-farm activities. In addition to farming and undertaking reproductive activities, women are actively engaged in micro-manufacturing enterprises and in trading agricultural and fishery products. In most developing countries, women's actual contribution to food security and the rural economy remains undervalued. They have less access to productive resources compared to their male counterparts. Quisumbing et al. (2014) stated that agriculture is underperforming because half of the farmers, women specifically, do not have equal access to resources and opportunities to engage in farm and non-farm activities. Empowered women who make decisions about planting materials and inputs have been found to be more productive in agriculture (Quisumbing et al. 2014).

Many food-production activities in Africa are attributable to women, making them the principal agents of food security and household welfare in rural areas. Women work as unpaid family labourers, self-employed producers, farm and non-farm employees, entrepreneurs, traders and providers of services, and technology researchers and developers (Hill 2011).

Women are largely responsible for production that benefits local consumption, including subsistence crops such as legumes and vegetables, operating on smaller plots and more marginal lands (Guendel 2009). While certain crops are often identified as women's or men's crops, gender roles vary, as do tasks, and in practice the divisions are blurred. For example, men may help to prepare the plots on which 'women's crops' are grown and women may be involved in weeding 'men's crops'. Because women lack access to capital and resources, their decisions about what they grow are more limited (Guendel 2009). For example, a study in rural Kenya showed that men were responsible for building the granary and women were responsible for most pre-harvest farm activities, including hand digging, harvesting, and transporting the crops. In another example, women farmers in Ghana chose to cultivate yams and cassava over maize—traditionally a 'man's crop' (Brody et al. 2014). This decision was made because these crops require fewer external inputs and are more labour intensive and are often referred to as 'women's crops'. These studies also indicate that decisions on what to plant and where to plant are mostly controlled by men, therefore sometimes excluding women from decision making (Guendel 2009).

Despite their vital contributions, women do not get credit for much of their farming labour, including laborious work such as weeding and post-harvest processing as well as food preparation, not to mention fuel and water collection and the myriad other household chores they perform which contribute directly to food availability and access.

Women's participation is also significant in activities such as planting and transplanting of crops, manual weeding, and harvesting. Male workers are generally engaged in land preparation, seedbed preparation, levelling, and care of irrigation canals. According to Quisumbing et al. (2014), women comprise about 43 per cent of the agricultural labour force in developing countries, ranging from 20 per cent in Latin America to 50 per cent in sub-Saharan Africa and East Asia. Hence women's contribution to labour as well as their managerial skills should not be underrated.

There is an increasing trend of FMFs in developing countries generally and sub-Saharan Africa specifically (World Bank 2007). In spite of this feminization, female farm managers in sub-Saharan Africa continue to face constraints that affect their ability to improve yield, profit, and efficiency in agriculture (World Bank 2007). These constraints include, *inter alia*: (1) women's legal and cultural status, which limits their degree of control on productive resources and inputs; (2) property rights and inheritance laws, which limit their rights to access land and other natural resources; (3) relationships among economic and ecological factors which include product market failures, poor marketing channels, seasonality of rainfall, and unavailability of fuel-wood; (4) exclusion from employment options due to skewed design and management of agricultural services such that less women are hired. Women are therefore major contributors to agriculture, and they play a prime role in ensuring the food security and nutrition status of their household members. It is conceivable that they could achieve much more in food production, provision, and utilization if agricultural researchers, plant scientists, extension agents, and policymakers could level the agricultural playing field (Pinstrup-Andersen et al. 1997).

Objectives

This chapter has three main objectives. First, the maize production trend across farm types is analysed. The second objective is to establish whether there are gender differences in maize technology use. The third objective is to examine the impact of NFI on agricultural investment. These objectives are examined in the context of farm type, defined by the gender of the farm manager. Previous studies (Kimenju and Tschirley 2009, Wambugu and Karugia 2014) have not adequately investigated these aspects along gender lines. The rest of this chapter is organized as follows. The next section describes the methods of data analysis, while the section following this discusses maize production and use of maize technology across different farm types. After this the impact of NFI on agricultural investment is discussed, followed by the concluding section.

Conceptual Framework and Hypothesis

The quality of planting material has a significant impact on crop productivity (Kibaara et al. 2009). The limited potential for further area expansion under maize cultivation due to diminishing availability of arable land implies that future growth in maize production would have to depend on yield gains made by widespread use of productivity-enhancing technologies, among which include hybrid maize varieties. This chapter looks at technology use and the impact of NFI on agricultural investment using a gender lens. Numerous studies have shown that, in terms of agricultural outputs, FMFs produce less than MMFs (Clark 2013, Palacios-López and López 2015). However, common knowledge of agriculture shows that the location and quality of farmland have a large effect on the level of output. Additionally, improved seed varieties, fertilizers, irrigation water, pesticides, extension services, and other technologies strongly affect agricultural output. NFI has also been shown to affect agricultural investment and therefore productivity. To truly understand the interrelationship between technology use, gender, and the impact of NFI on agricultural investment, these factors require empirical analysis. Figure 10.1 is a simplified conceptual model that tries to depict the interrelationship among these factors. The plus sign indicates the expected positive relationship among the technologies and agricultural yield.

Farm households often have to make complex decisions regarding consumption, investment, and income-earning activities. These decisions are influenced by a variety of external and internal factors. Farmers are often expected to invest NFI in farming if the farm investment allows them to maintain or increase farm output (Harris et al. 2010). Investing part of the NFI in the farm in this case is expected to increase agricultural productivity.

The working hypothesis in this chapter states that MMFs and FMFs are equally as productive in Kenya's agricultural sector. However, agricultural output will differ by gender for a variety of reasons.

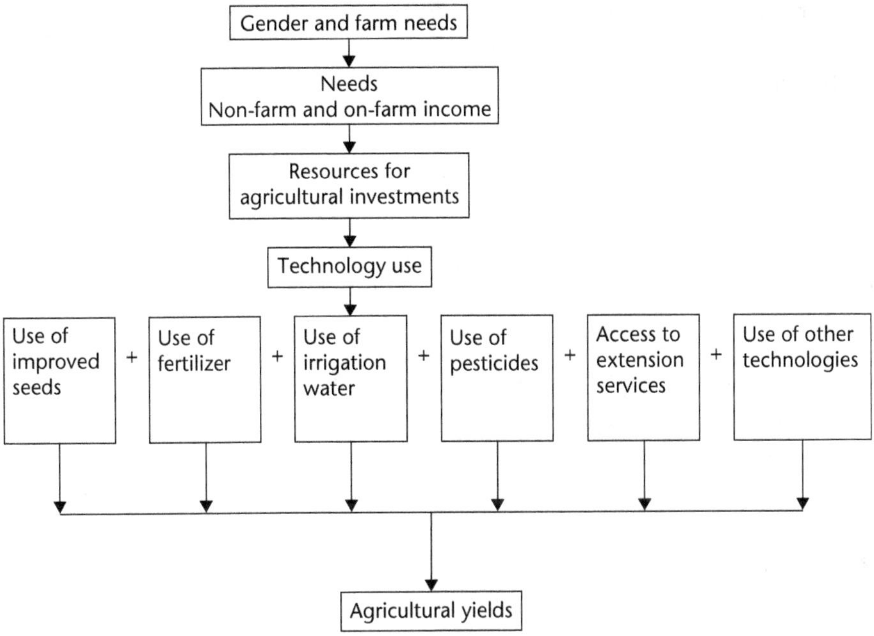

Figure 10.1 A simplified conceptual model.
Source: Adapted and modified from Clark (2013).

The chapter argues that technology use and investment of NFI in agriculture can explain differences in maize productivity. As FMFs are more likely to be relegated to lower ranks in terms of access to agricultural inputs and NFI, their productivity will tend to suffer. As already established by other studies (Ahuja 1998, López 1998, Ariga et al. 2006, Kibaara et al. 2009, Ogada et al. 2010), the chapter anticipates that access to inputs such as fertilizer and improved seeds will have a significant effect on maize production. The analysis also anticipates that use of NFI will enhance agricultural investment. Thus investment in farm inputs might be affected by access to NFI. Farmers use both on-farm and non-farm income for farm investment in order to enhance productivity and reduce drudgery. NFI is necessary to complement on-farm income for farm investment up to the required levels depending on the farmer's total income and priorities. This chapter zeroes in on the impact of NFI on fertilizer use, for it is a key agricultural input.

Data and Methodology

This chapter uses panel data collected during three rounds (Afrint I, II, and III) from ten villages in two regions (counties) of Kenya (Nyeri and Kakamega).

Data that was collected included *inter alia* maize production, use of technology on maize, and sources of NFI. These data were disaggregated by gender of the farm manager.

Methods of Data Analysis

In order to examine maize production trends and associated technology use, descriptive statistics were used. In particular, means, percentages, and proportions were used. In addition, the double-hurdle model and marginal effects were used to assess the impact of NFI on agricultural investment. The analysis assesses how engagement in non-farm work affects the decision to invest. Double-hurdle models were run for fertilizer demand in maize production. The models were disaggregated by the major sources of NFI and aggregated for the total NFI.

Results and Discussion

Maize Production Trends

This section analyses trends on maize productivity using household panel survey data collected from 300 households in ten villages in the two regions. This panel data analysis overcomes problems of comparability and differences in sample design that compromise other trend assessments and thus provides a unique opportunity to evaluate changes in smallholder agricultural productivity.

Results for maize productivity analysis are presented in Table 10.1. The average maize yields for the two regions show no consistent trend. For Kakamega the yields increased from 819.1 kg/ha during Afrint I to 1,459.4 kg/ha during Afrint II, before dropping to 1,104.5 kg/ha during Afrint III. The Nyeri case is the same, with the yields increasing from 257.3 kg/ha during Afrint I to 1,101.1 kg/ha during Afrint II, and then dropping to 349.9 kg/ha during Afrint III. The impressive increase during Afrint II can be attributed to

Table 10.1. Maize yields (kg/ha) by gender and region

	Afrint I			Afrint II			Afrint III		
	Kakamega	Nyeri	Sig.	Kakamega	Nyeri	Sig.	Kakamega	Nyeri	Sig.
FMFs	831	199	***	1,106	1,263		1,119	474	***
MMFs	807	315	***	1,813	940		1,090	316	***
MMFs – FMFs	–24	116		707	–323		–28	–158	
Sig. (MMFs – FMFs)		***							
Mean production	819	257	***	1,459	1,101		1,105	395	***

the resumption of government input subsidies. However, the drop during Afrint III can be accounted for by farmers diversifying into other crops. In Nyeri there has been a diversification towards the growing of horticultural crops which fetch better prices than maize. In Kakamega, farmers are diversifying into the growing of sugarcane. Similar findings have been reported by the Ministry of Agriculture (Kibaara et al. 2009).

Nationally the rising maize yield during the period between Afrint I and II is attributed to a combination of good weather, use of improved seeds, higher fertilizer application, and adoption of modern farming techniques and technologies (Republic of Kenya 2008). Kakamega recorded higher average yields throughout the panel period compared to Nyeri. This is because Kakamega has better agro-ecological potential and some of the villages sampled (Mukuyu and Munyuki in particular) are areas where maize is predominantly produced under monoculture cropping. Disaggregating maize yield by gender of the farm manager, the results show a consistent upward trend for FMFs in Kakamega while for MMFs the trend is mixed. In the case of Nyeri, results for both FMFs and MMFs show a mixed trend with an increase during Afrint II and a drop during Afrint III.

The impressive growth in maize yields, especially for Kakamega FMFs and during Afrint II, could be attributed to several factors, including increased input use as well as physical and market infrastructural developments. However, data presented in Table 10.1 do not show any significant difference in maize yields between MMFs and FMFs save for Nyeri during Afrint I and II. This finding is therefore not in tandem with other studies (Clark 2013, Palacios-López and López 2015) that have reported significant differences in agricultural productivity between MMFs and FMFs. This result can possibly be attributed to the various government affirmative action initiatives aimed at gender equity in all sectors of the economy.

The results in Table 10.1 show significant differences in maize yields in Kakamega between Afrint I and II and not between Afrint II and III. The significant difference can be attributed to the resumption of the government input subsidy programme and other policies aimed at revitalizing agriculture instituted immediately after the National Rainbow Coalition Government came into power in 2002. However, Nyeri doesn't show any significant difference during the panel period. This can be explained by the fact that Nyeri is traditionally not a maize-growing area—most of the farmers grow cash crops (coffee, tea, and horticultural crops).

Technology Use in Maize

This section looks at technology use in maize in both regions. This is disaggregated by gender of farm manager to try and understand gender differences

Table 10.2. Technology adoption for maize by survey round and gender (%)

Type of technology		Afrint I			Afrint II			Afrint III		
		MMFs	FMFs	Sig.	MMFs	FMFs	Sig.	MMFs	FMFs	Sig.
Use of hybrid maize	Kakamega	39.3	22.7	***	53.3	25.3	***	54	33.3	***
Varieties	Nyeri	48	41.3	**	2.0	4.7	***	2.0	0	***
Use of chemical fertilizers	Kakamega	48.7	28.7		44	22.7	*	50	32	
	Nyeri	41.2	34.5		43.2	28.4		55.4	25.7	
Use of pesticides	Kakamega	5.33	3.33	*	6.00	5.33		14.7	5.33	***
	Nyeri	2.70	0.67		12.84	6.67		39.13	17	
Use of hoe	Kakamega	24.7	16.7		32	20		32.7	12	
	Nyeri	50.7	32.4		53.4	31.1		52.0	46.6	
Use of oxen plough	Kakamega	32.0	20.7		32.7	12	**	32.7	12	*
	Nyeri	12.2	5.41	***	11.9	0.0	*	0.68	0	***
Use of tractor plough	Kakamega	4.0	2.0	**	2.0	0.67		10	4	
	Nyeri	0.68	0.0	***	3.4	2.0	*	0.68	0.0	

in technology use. The technologies applied in maize production are depicted in Table 10.2.

The results in Table 10.2 show a general increasing trend in the proportion of farms planting hybrid maize varieties over the panel period, from 39.3 per cent during Afrint I, through 53.3 per cent in Afrint II, to 54.0 per cent in Afrint III for MMFs in Kakamega. The trend is the same for FMFs in Kakamega, with the percentages increasing from 22.7 per cent, through 25.3 per cent, to 33.3 per cent during Afrint I, II, and III respectively. However, analysis for Nyeri reveals a decline for MMFs from 28 per cent using hybrid maize during Afrint I to 2 per cent during Afrint II and III. For FMFs adoption dropped from 41.3 per cent during Afrint I, to 4.7 per cent and 0 per cent during Afrint II and III respectively. The declining use of hybrid maize varieties in Nyeri over the panel period could be attributed to some unscrupulous businessmen and extension workers who provide farmers with uncertified seeds. While hybrid maize varieties contribute towards improved crop yields, their use must be supplemented by other productivity-enhancing inputs, mainly fertilizer, to exploit their full productivity potential. In both regions, it is observed that there is a significant difference in the adoption of hybrid seed between MMFs and FMFs.

Fertilizer use for maize production in both FMFs and MMFs show no consistent trend (Table 10.2). In Kakamega, fertilizer usage decreased from 48.7 per cent during Afrint I, to 44 per cent during Afrint II, before rising to 50 per cent in Afrint III among MMFs. For FMFs, the same trend is observed where we see a decrease from 28.7 per cent to 22.7 per cent, and then a rise to 32 per cent during Afrint I, II, and III respectively. In Nyeri we see a consistent increase in fertilizer usage from 41.2 per cent during Afrint I, through 43.2 per cent during Afrint II, to 55.4 per cent during Afrint III for MMFs. Taking cognizance of the declining trend in the percentage of farmers adopting hybrid maize and the consistent increase in fertilizer usage, it is only logical to argue that fertilizer is used in growing other crops. However, for FMFs there is a consistent decrease in fertilizer usage from 34.5 per cent, through 28.4 per cent, to 25.7 per cent during Afrint I, II, and III respectively. This can be attributed to FMFs resorting to other soil fertility-enhancing technologies such as use of farmyard manure. In both Nyeri and Kakamega the proportion of FMFs using fertilizers is lower than that of MMFs, but the difference was not significant.

Though there has been a general increase in smallholder fertilizer use in Kenya over the past decade (Kibaara et al. 2009), results from this study do not mirror the same trend, especially for FMFs in Nyeri. This can be attributed to the gender-linked differences in access to inputs. However, according to Kibaara et al. (2009), the general increase in fertilizer usage, especially among the MMFs, may be attributed to several factors, namely: (1) increased access of fertilizer by smallholder farmers due to availability of the input in smaller packs that more farmers can afford; (2) reduction in the distance from the household to the nearest fertilizer stockist, reflecting increased investment in private fertilizer retailing; (3) a reduction in real fertilizer prices in Kenya up to 2007, reflecting reduced fertilizer marketing costs (Ariga et al. 2006)—this trend was reversed since 2007 following the dramatic rise in world fertilizer prices; and (4) more farmers have been organized into groups, providing a variety of benefits such as group loans for input purchase, information to improve farmers management practices such as soil testing services, increased awareness of the role of fertilizers in increasing maize productivity, and information on how and when to apply fertilizer efficiently.

Analysing patterns in simultaneous use of fertilizer and high-yielding varieties (HYVs) of maize can shed more light on the observed productivity trends and provide information that can be useful in proposing measures to improve agricultural productivity. Table 10.3 presents the results of this analysis.

Table 10.3 shows a consistent general increase in the proportion of farms combining fertilizer and HYVs for maize in Kakamega for both MMFs and FMFs and for the panel period. However, the trend for Nyeri is mixed where we see a decline during Afrint II and an upward trend during Afrint III for MMFs

Table 10.3. Percentage of farmers using both fertilizers and high-yielding maize varieties

	Afrint I		Afrint II		Afrint III	
	Kakamega	Nyeri	Kakamega	Nyeri	Kakamega	Nyeri
MMFs	34.34	33.1	36.15	22.1	38.83	28.2
FMFs	20.03	27.58	18.18	15.73	24.66	12.9
Sig.	**		***	**	**	***

and a general decline for FMFs. The results show no significant difference in the proportion of MMFs and FMFs that have adopted both fertilizer and high-yielding maize varieties for Afrint I in Kakamega and Afrint II and III in Nyeri. These results are not in agreement with the work by Kibaara et al. (2009), who found that regionally there are distinct variations in the rate of adoption of the combined fertilizer-improved seed package. Adoption of new maize varieties and enhanced availability of HYVs has been fuelled by the liberalization of the seed market. Distance to market, both for inputs and outputs, has been a key issue in productivity improvement (Omamo 1998).

Application of pesticides is another important technology used in maize production and for reducing post-harvest losses. Trends in pesticide use (Table 10.2) on maize show a consistent general increase in Kakamega for both MMFs and FMFs, though the trend stagnates for FMFs during Afrint II and III. In Nyeri there is a consistent upward trend in the use of pesticides for both MMFs and FMFs, which can be attributed to extension services (provided by the public and private sectors) where farmers have been sensitized on the role of pesticides in controlling pests and thereby boosting production, but which also carries significant risks to human health. Sales of pesticides to developing countries, while remaining a relatively small proportion of global turnover (about 25 per cent), have increased significantly in some developing countries (London et al. 2002). Ironically, this is occurring at a time when many developed countries have begun to implement pesticide-reduction programmes in the belief that excessive use of pesticides is harmful to workers, consumers, and the environment.

Small-scale farmers use either the hoe, ox plough, or tractor plough to prepare land. Table 10.2 shows that for MMFs in Kakamega the use of the hoe is on the increase, with the proportion of farmers using it rising from 24.7 per cent, through 32 per cent, to 32.7 per cent over the course of the three Afrint rounds. This can be accounted for by the decreasing farm sizes and the high costs associated with the usage of the ox and tractor plough. However, for FMFs there was a marginal increase from 16.7 per cent during Afrint I to 20 per cent in Afrint II. This however decreased to 12 per cent during Afrint III. In Nyeri, usage of the hoe shows mixed results. For MMFs there is an increase from 50.7 per cent to 53.4 per cent, then a marginal decrease to 52 per cent during Afrint I, II, and III respectively. For FMFs the use of the hoe shows mixed results, with a decrease from 32.4 per cent to 31.1 per cent during Afrint I and II, and then an increase to 46.6 per cent during Afrint III. The results also show that for MMFs in Kakamega, the use of the ox plough increased marginally from 32 per cent to 32.7 per cent, before stagnating at 32.7 per cent in Afrint I, II, and III respectively.

For FMFs the same trend is maintained. There is a decrease from 20.7 per cent during Afrint I to 12 per cent during Afrint II. Stagnation at 12 per cent is observed for Afrint II and III. There is a consistent decrease in the usage of the ox plough for MMFs from 12.2 per cent, through 11.9 per cent, to 0.68 per cent during Afrint I, II, and III respectively. For FMFs the usage of the ox plough is very minimal. During Afrint I only a paltry 5.41 per cent used the ox plough, while in Afrint II and III none of the FMFs used the ox plough. Use of the tractor plough is not a common or popular method of land preparation. In MMFs the percentage usage declined from 4 per cent during Afrint I to 2 per cent during Afrint II, however it increased to 10 per cent during Afrint III. For FMFs, use of the tractor plough decreased from 2 per cent to 0.67 per cent during Afrint I and II respectively, before rising to 4 per cent during Afrint III.

In Nyeri, tractor usage was extremely low at 0.68 per cent during Afrint I, rising to 3.4 per cent during Afrint II, before sliding back to 0.68 per cent during Afrint III. For FMFs in Nyeri, the use of the tractor plough is at 0 per cent for Afrint I and Afrint III, while a small percentage of 2 per cent used it during Afrint II. The low usage of land-preparation implements, especially among FMFs, could be a pointer to the need to develop gender-friendly agricultural technologies. A number of studies (IFAD 1998, FAO 2011) have concluded that no quick-fix is possible for improving the production tools and implements used by FMFs in Africa. However, over time, and with appropriate action by governments, development agencies, NGOs, and the private sector, the situation could be improved. The constraints FMFs face (such as access, costs, and unfriendliness) in using farm implements and tools, and actions necessary to overcome them, are issues of research and policy.

In a nutshell, the results presented in this section confirm the widely accepted observation that technology adoption and use is positively correlated with maize productivity. This is consistent with the findings by López (1998), Ahuja (1998), and Ogada (2010). However, to enhance productivity and to reduce drudgery farmers use both farm and non-farm income in their farm investment. NFI is necessary to complement farm income for farm investment. The next section therefore addresses the impact of NFI on agricultural investment, zeroing in on fertilizer which is a key agricultural input and which is often used to complement improved seed.

Investment of NFI in Agriculture

This section considers factors that determine the amount spent on fertilizer (a key agricultural input) in maize (the main staple crop) production using the double-estimation technique. The section uses data collected during the Afrint III period. As a robustness check, the estimated parameters are compared to the corresponding standard Tobit estimation, which is defined as

$$t_1^* = X_i'\beta + \epsilon_i \quad \text{with} \quad \epsilon_i \sim N(0, \sigma^2) \quad \text{and} \quad i = 1, \ldots, n \tag{10.1}$$

$$t_i = \begin{cases} t_i^* & if\ t_i^* > 0 \\ 0 & if\ {}_t^{i*} \leq 0 \end{cases} \tag{10.2}$$

Where t_i^* is a latent endogenous variable representing individual i's desired level of expenditure on fertilizer, and t_i is the corresponding actual observed expenditure on fertilizer, X_i is a set of individual characteristics that explains the use and level of expenditure on fertilizer, and β is a corresponding vector of parameters to be estimated, ϵ_1 is assumed to be a homoscedastic, normally distributed error term.

In the double-hurdle model specification an individual has to overcome two hurdles in order to report a positive amount of money spent. The first hurdle is based on whether the farmer uses fertilizer in maize production, and the second hurdle models the decision on how much to invest in the fertilizer. The double-hurdle model, originally formulated by Cragg (1971) modifying the standard Tobit model, assumes that two hurdles are involved in the process of investment decisions, each of which can be determined by a different set of explanatory variables. A different latent variable is used to model each decision process,

$$y_{i1}^* = w_i'\alpha + v_i \quad \text{Investment decision} \tag{10.3}$$

$$y_{i2}^* = x_i'\beta + u_i \quad \text{Level of investment} \tag{10.4}$$

$$y_i = x_i'\beta + u_i \text{ if } y_{i1}^* > 0 \text{ and } y_{i2}^* > 0$$

$$y_i = 0 \quad \text{Otherwise} \tag{10.5}$$

We can envision simultaneity (e.g. use of fertilizer and hybrid seed). However, as Table 10.3 depicts, the majority of the farmers do not use both. We can also envision multicollinearity (e.g. between farm income and NFI) of some of the variables used in the model. Correlation analysis showed no multicollinearity. The variables used in the model were those that were thought to be important in influencing the probability and intensity of fertilizer use in maize.

Tables 10.4 and 10.5 present parameter estimates of the fertilizer demand model using Afrint III data. Results are aggregated for NFI (Table 10.4) to first understand the impact of total NFI, and then disaggregated by source of NFI (Table 10.5) to understand which of the NFI sources are important in farm investment. The dependent variable is the total amount of financial resources spent on fertilizer per hectare of maize grown. Coefficients in the first hurdle indicate how a given decision variable affects the likelihood (probability) of adopting fertilizer in maize. Those in the second hurdle indicate how decision

Table 10.4. Probability of investing in fertilizer and the intensity of use in maize (aggregated non-farm income)

Variable	First hurdle	Sig.	Second hurdle	Sig
Sex of the household head	0.191		1,298.1	
Age of the household head	0.016	**	−1.3722	
Education	0.051	**	−14.377	
Distance to the nearest town	0.014		−242.99	
Maize area recent season	−0.031		8,793.1	*
Plan to sell maize	0.783	*	8,802.3	*
Access to agricultural credit	1.098	*	−48.28	**
Agricultural income	−8.8E-07		0.005	
Non-farm income	−1.3E-06		−0.002	
Hybrid seed	0.633	**	6,037.0	
Maize production previous season	3.2E-04		2.914	*
Constant	−1.465	*	−21,122	*
Log likelihood	−2,340.8			
Wald χ^2	42.96			

Table 10.5. Probability of investing and the intensity of fertilizer use in maize (disaggregated non-farm income)

Variable	First hurdle	Sig.	Second hurdle	Sig.
Sex of the household head	0.22		1,209.0	
Age of the household head	0.015	**	34.20	
Education	0.058	**	−10.74	
Distance to the nearest town	−0.015		−192.8	
Maize area recent season	−0.138		8,623.2	*
Plan to sell maize	0.815		8,323.3	*
Access to agricultural credit	1.097	*	483.7	
Agricultural income	−1.1E-06		0.008	
Hybrid seed	0.620	**	5,954.8	
Maize production previous season	4.0E-04		3.066	*
Salary	−1.7E-06		0.002	
Micro-business	1.4E-06		−0.004	
Large-scale business	1.7E-05		−0.001	
Rent/interest	−6.9E-06		0.020	
Pension	−3.8E-06		0.003	
Remittances	1.0E-05		−0.085	
Constant	−1.418		−22,502.2	
Log likelihood	2,336.5			
Wald χ^2	44.76			

variables influence the amount spent on fertilizer per hectare. Marginal effects depict changes in the probability of adoption of fertilizer for an additional unit increase in the independent/decision variable.

Results in the first hurdle indicate that age of the household head, education, plan to sell maize, access to agricultural credit, and hybrid seed are statistically significant decision variables that influence the probability of adopting fertilizer technology among farming households in the study areas.

The probability of fertilizer adoption increases by 1.6 per cent for every one year increase in the age of the household head. This implies that as age increases, the probability of adopting fertilizer increases.

A unit increase in education level increases the likelihood of adopting fertilizer by 5.1 per cent. The farmers' age and education level had positive and significant coefficients. This indicates that the probability of investment in fertilizer increases with age and as the farmers gained more experience in farming. This might suggest that older and more experienced farmers may be using NFIs to finance farm investments or substitute higher NFI for farm income. This could be attributed to the experience gathered over the years in coping with the menace of soil infertility. However, sex of the farm manager did not significantly influence the decision to invest in fertilizer.

The plan to sell maize increases the probability of fertilizer adoption by 78.3 per cent. This means that commercialized farmers have a very high probability of adopting fertilizer use in maize in the belief that the more fertilizer they use the more maize they can harvest for selling.

The use of hybrid seed increases the probability of fertilizer adoption by 63.3 per cent, and was a significant factor influencing the probability of investment in fertilizer. Households using hybrid seed also tend to use fertilizer, and thus the two inputs are likely complements.

The results further reveal that access to agricultural credit increases the probability of fertilizer adoption by 109.8 per cent. This implies that access to agricultural credit is a very important factor in overcoming constraints to agricultural technology adoption. Agricultural credit services are the major sources of finance to those farmers who adopt improved agricultural technologies like fertilizer application (Kimenju and Tschirley 2009). Although agricultural credit is mostly provided for cash crop farming, there is some expected spillover effect to cereals and other food crops. It is therefore expected that households that can access agricultural credit will have a higher likelihood of using fertilizer. Access to agricultural credit had the expected positive and significant effect on the decision to invest in fertilizer.

Distance to the village centre was included to proxy for transport costs. Proximity of farmers to markets is essential for timely input delivery and output disposal, and results in inputs and outputs incurring less transport costs. The coefficient of distance was, however, not significant for the intensity of fertilizer use, meaning farmers interested in using fertilizer were not deterred by cost of transport.

The previous season's maize production was included based on the expectation that the amount of maize harvested the previous season positively influenced the intensity of investment in fertilizer, meaning that when farmers experienced increased production they tended to invest more in fertilizer in the following season. However, this variable was found not to be significant.

Results further show that NFI has no significance on the probability of adopting fertilizer technology. This could be attributed to farmers investing their NFI in less risky and perhaps more lucrative non-farm activities. NFI had a negative but insignificant coefficient for adoption and intensity models. The negative and insignificant impact of NFI imply that, holding other factors constant, NFI seems not to impact both adoption and intensity of investment in fertilizer. This suggests that these households are not using some of their non-farm earnings to purchase fertilizer for maize production. In this case, non-farm earnings may not be needed to relieve cash constraints for fertilizer purchase. Therefore, it can be concluded that NFI is not driving the level of farm investments.

Table 10.5 presents results with disaggregated NFI. This analysis was done to identify which of the different types of NFI may be driving decisions to invest on the farm.

Results from the first hurdle reveal that none of the NFI activities were significant in influencing the probability of adopting fertilizer technology. Micro-business, large-scale business, and remittances had very small increases in the probability of fertilizer adoption, while salary, rent/interest, and pension tended to very marginally decrease the probability of fertilizer adoption. Income from micro-businesses, large-scale businesses, and remittances had negative and insignificant small coefficients in the second hurdle, implying that income from these activities was not important in determining the level of investment in fertilizer.

Salary, rent/interest, and pension had positive and insignificant impacts on the level of investment in fertilizer. The very small positive and insignificant coefficients suggest that for households using income from these sources, the level of investment increases very slightly as the income increased. Income from these sources are likely to be in high amounts and on a regular basis, and therefore farmers might not see the need to invest so much of it on the farm. These incomes might also make it possible to invest in non-farm activities. These results corroborate the results in Table 10.4 that NFI is not significantly influencing farm investment.

Conclusions and Policy Implications

In this chapter, we have examined aspects of maize productivity, technology use in maize, and the impact of NFI on agricultural investment, giving them a gender dimension. The study first concludes that there are no significant differences in maize yields between MMFs and FMFs in the study area. Second, technology use for maize production was lower and significant in some instances for FMFs. Significant differences, especially in the use of hybrid seeds and the tractor plough, were noted. A third conclusion is that the NFI is not used in farm investment. The results of the double-hurdle models showed that NFI had no significant effect on adoption and intensity of fertilizer use.

One policy recommendation is that government policies, programmes, and interventions targeting improved agricultural production should continue considering FMFs and cushion them against risks associated with technology adoption. The chapter has also presented evidence to show that NFI is not invested in agriculture. Consequently, another policy implication is that policies that increase both on-farm and non-farm income for households should be encouraged, given the complementary roles that they play. Any entry barriers for disadvantaged households, especially for the FMFs, to participate in higher-paying non-farm activities need to be overcome. In addition to increasing household income directly, improved access to non-farm activities can also lead to positive indirect effects. Especially when rural financial markets are imperfect, cash from NFI can partly be invested in agriculture, thus also increasing farm production and income.

Given the complementarities between non-farm and farm income, and the fact that both sectors actually face similar constraints, appropriate policy instruments can actually serve both purposes. For instance, accessible credit schemes can facilitate the establishment of non-farm businesses and promote agricultural development simultaneously. The argument here is that there are a lot of synergies and positive spillover effects between agricultural and non-agricultural sectors. Over time, it is likely that the relative importance of the non-farm sector will increase further. Improved opportunities in rural areas targeting FMFs and other disadvantaged groups can help level the playing field and reduce the gender gap, thereby improving their well-being.

Acknowledgements

The authors would like to thank Dr Lucy Ngare, Rosaline Karimi, Tetu Mwenda, field enumerators, and farmers for their assistance during data collection and analysis.

11

Agriculture, Diversification, and Gender in Rural Africa: What Lessons Can We Learn?

Agnes Andersson Djurfeldt, Fred Mawunyo Dzanku, and Aida Cuthbert Isinika

As noted in Chapter 1, as well as in several other chapters, numerous changes have characterized the decade and a half over which the Afrint project has been running. In some respects, the original aim of the project—to focus on small farmers as the prospective bearers of agricultural intensification—has been outpaced by policies and events that have resurrected small farms as the linchpin of rural development and poverty reduction. In this sense, the smallholder focus of the Maputo Declaration and the *World Development Report 2008* (World Bank 2007) for was in some ways strengthened by the renewed emphasis on domestic agriculture following the global food price crisis of 2008. A spate of policies focusing on smallholders has characterized the project period.

More recently, however, the unfolding social and economic realities *outside* agriculture are providing opportunities as well as challenges for the smallholder model. Such dynamics—described in detail in earlier chapters—include demand generated by the relatively rapid growth of the non-farm sector, tempered in some respects by the nature of African urbanization and a youthful demographic bulge of labour market entrants. At the same time, signs of growing socio-economic differentiation *within* agriculture along the lines of gender and generation are apparent. Concerns over land fragmentation and the fear that the relatively limited prospects of women, migrants, and other marginal groups will dwindle further over time are expressed by researchers as well as policymakers. The prospects for leaving agriculture altogether appear dim for the rural residents of sub-Saharan Africa, however, as suggested by several studies showing the persistently high share of income generated from agriculture.

The project period (2002–15) thus overlaps with a rising interest in smallholder-based agriculture among researchers and policymakers, but also an increasing recognition of the challenges facing the smallholder sector and particular groups of small farmers. More recently, however, policy experiments to encourage emergent or middle sized farmers, in combination with more longstanding support for large-scale agriculture in several of the countries under study, could portend the gradual abandonment of the smallholder model. Increasingly politicized land deals and speculation do not augur well in this respect.

This book set out with three aims, the simplest one relating to summarizing and analysing patterns of change among the households followed by the project over close to fifteen years, departing from the overarching themes of the research: agriculture, rural livelihoods, and gender. This largely descriptive aim was complemented by a theoretical one—aspiring to nuance macro-level narratives of structural change through analysing local-level realities of rural livelihoods that straddle the farm and non-farm sectors, as seen in sometimes localized specifics. Finally, we envisaged a practical outcome to which the empirical and analytical aims would contribute, namely a policy-related ambition to outline alternative options to current policy practices.

Agriculture, Rural Livelihoods, and Gender: A Longitudinal Perspective

So, what do the data tell us about the smallholder sector and rural livelihoods over the past fifteen years? We identify a set of trends for the sample as a whole, which broadly speaking can be summarized as increasing dynamism accompanied by rising polarization. The changing relationship between the farm and non-farm sectors and the role of the state in steering and sometimes misguiding agricultural development are other tendencies that emerge from the data.

Differentiation Amid Agricultural Dynamism

The growing vigour of smallholder agriculture is evinced in the cross-sectional data through a set of positive trends: on average, farm sizes have increased; production has risen substantially for grain crops; crop diversification away from grain staples towards more lucrative food crops has occurred widely; while commercialization has generally increased, especially for maize and rice. The lion's share of farmers was commercialized at the time of Afrint III, with the share of commercialized farmers increasing from 71 per cent to 84 per cent during the course of the study. At the same time, yields have remained largely stagnant—production increases have therefore been tied to increasing areas under cultivation—and yield gaps remain stubbornly persistent over the project phases. The distribution of land, moreover, is becoming increasingly unequal—while the land sizes are stagnant at the bottom of the distribution they have increased considerably at the top.

The gender dynamics of land access show that households headed by women have persistently smaller cultivated areas, but also that changes in land sizes since the first phase of the project have affected these households negatively when compared with their male counterparts. In regions where mean land size increased, the gender gap between female- and male-headed households has widened, even if land sizes for the former increased on average. Slight increases in formalization of tenure can be noted over the period, and where this has occurred it has also been in favour of male-headed households. Control over land was, however, generally gender neutral.

Overall, female farm managers have poorer access to agricultural assets such as land, water sources, and livestock. In combination with lower access to male labour specifically, this reduces the possibilities for producing marketable surpluses on these farms. While it must be stressed that commercialization patterns shift over time and that patterns vary nationally, the strong trends towards commercialization in the maize sector in Zambia has been accompanied by a growing differentiation based on gender. Markets for non-staple food crops have also been masculinized in Tanzania, Malawi, and Mozambique, while they have been feminized in the case of Kenya. In terms of cash incomes, gender gaps have widened and also emerged in some countries since the second round of data collection. Four of the six countries had statistically significant gender gaps in cash income by the time of Afrint III, compared with two for the earlier period. Such gaps are tied predominantly to farm-based income sources, where better access to agrarian resources among male farm managers translates into higher incomes.

New Patterns of Diversification

While agricultural sources dominate the income portfolios of the respondents—indeed, on average more than two-thirds of the cash income generated is based on agriculture—households specializing in agriculture have lower incomes when compared with diversified households. Despite cultivating larger plots of land, using fertilizer seed technology, and being highly commercialized, earning non-farm income (NFI) has a stronger differentiating influence than superior access to agrarian resources.

Whether NFI can in this way compensate for poorer access to agrarian resources among female-headed households is less clear, however—indeed, the only income source that is persistently pitched towards female-headed households is remittances. Regionally, however there are exceptions in this respect, with micro-businesses complementing male incomes for female-headed households and for women within male-headed households in some areas. Nonetheless, the principal indication that emerges from the data are the relatively limited possibilities that women have for generating cash incomes in general—whether from agriculture or otherwise, whether as female heads of household or as women in male-headed households.

While gender gaps in assets as well as incomes have widened over the period, female-headed households have nonetheless benefited from the dynamism that has characterized the smallholder sector in some countries and regions, especially since Afrint II. The housing standard has increased across the board, and here there are no differences based on the sex of head of household. Similarly, ownership of mobile phones has expanded dramatically, and again there are no gender-based differences. In contrast, the ability to save and the transportation assets owned show the familiar tendencies of improvements accompanied by growing gender gaps. In general, therefore, the patterns in the cross-sectional data point to rising living standards, but also growing differentiation based on gender.

The cross-sectional findings for the whole dataset are to some extent also corroborated in the chapters using panel data. The three key processes of structural transformation (ST) studied in Chapter 5 show that for the second panel period, grain intensification was a gender-neutral process, whereas commercial diversification into markets for non-staple crops was skewed against female farm managers. In turn this suggests that more lucrative commercial opportunities have opened up outside the staple crop sector, but that these are predominantly seized by male-headed households. While intensification in food grains may be benefitting female farm managers and their families through improved food security, commercialization in more lucrative food crops favours male-headed households. Conversely, non-farm diversification does not discriminate against female-headed households, and is also negatively related to increases in farm sizes and positively connected with grain intensification. Moreover, the process is more common in villages with unequal land distribution, while it is also negatively related to being part of the upper decile of landowners at the village level. In turn this seems to suggest that livelihood diversification may be a more forthcoming option for households that have smaller agrarian assets, such as female-headed households. As suggested by the cross-sectional data, non-farm diversification is on the whole associated with higher incomes than specialization within farming. Results from the panel data also suggest a shift from distress-based diversification towards more accumulation-based processes. While commercial dynamism within agriculture since the second round of data collection has favoured male farm managers, movement into the non-farm sector during the same time has been beneficial for both groups.

While the dataset as a whole suggests a complementary role for non-farm diversification and grain intensification, the country-level study on Ghana (Chapter 9) points to a slightly different pattern, and reinforces the need for regional contextualization of linkages between farm and non-farm sectors. In the region with poorer agricultural potential, increasing non-farm earnings are connected to lower agricultural labour productivity. This relationship is found only when male earners generate such income, however. For the more dynamic region, farm labour productivity drops with female generated NFIs, but rises when men earn NFIs, pointing to a competing role between female non-farm earnings and agriculture, but a complementary one between male-earned NFI and agriculture in regions of relatively high agricultural potential. While non-farm diversification may benefit the household in terms of higher incomes and less vulnerability to seasonality and price volatility, such incomes may therefore not always be related to higher agricultural productivity.

Agricultural Policy

Several of the contributions point to the influential role of state policies. Many of the countries under study have experimented with input subsidies targeted at smallholders, but also more recently at the high-end segment of the sector—as shown in the comparative discussion on Zambia and Malawi (Chapter 8). The country-level experiences vary in this regard, with agricultural subsidies in Malawi, for instance, failing to make any long-term improvements in food security and self-sufficiency (Chapter 7). Here other small-scale initiatives appear to have been more fruitful in terms of encouraging diversification in the study sites, although the scaling up of such projects pose considerable challenges of their own. Nonetheless, as suggested by the findings from Tanzania, the on-and-off nature of many subsidy schemes, and their concomitant effects on smallholder production and productivity, point to the need for sustainable policies—whether based on subsidized inputs or not (Chapter 6). Moreover, several other types of rural development policies related to improving accessibility and lowering transaction costs may be equally relevant.

Theoretical Implications

Standard models of ST depict a combination of processes which together interact to move economies as well as people out of agriculture and rural areas and into non-farm pursuits in urban areas—at least at the macro level. The most fundamental of these processes relates to raising agricultural productivity in staple crops such as grains, roots, and tubers. Standard models of ST rest on a number of presumptions in this regard—suppositions based on the historical experience of the countries of the Global North. On the one hand, the exit from rural areas constituted a clean break with agriculture in the sense that livelihoods shifted into urban activities, and on the other, such activities were predominantly centred on manufacturing industries that developed during the Industrial Revolution. Several caveats can be raised with respect to present-day Africa in relation to both of these aspects: on the one hand, as suggested by evidence from Africa as well as other parts of the Global South, the break with agriculture is not quite as definite as presumed—instead, as suggested also by our findings, livelihoods increasingly straddle the farm and non-farm sectors as well as urban and rural spaces (Andersson Djurfeldt 2014, Andersson Djurfeldt 2015). One of the reasons for this pattern relates to the nature of African urbanization, driven as it is by consumption, natural resource extraction, and trade, rather than manufacturing (McGranahan et al. 2009, Gollin et al. 2016). The other is connected to the limited possibilities for raising smallholder productivity, as suggested by the persistent yield gaps found in the study sites. Rather, what we seem to be seeing is a process of grain production increases based on expansion in some regions and an intensification based on value rather than volume in other regions. In turn, this may call for a renegotiation of ST narratives to acknowledge spatial differentiation, as well as the different pathways of agrarian transformation that Africa may take in the future.

Overall, the gender aspects have been largely absent from discussions of agrarian change over the long term, possibly because the family structure of the historical Global North, as well as present-day South East Asia, is different from sub-Saharan Africa. Given the relatively large share of female-headed households in the dataset and the particularities of women's exclusion from agrarian resources, an understanding of the gendered patterns of agrarian transformation as well as the linkages to the non-farm sector is overdue. In this regard, somewhat surprisingly we show that grain intensification is largely gender neutral and that yield gaps in grain production based on sex are small. Nonetheless, with respect to the two processes described above (expansion-based increases in grain production and intensification through high-value crops), female-headed households, although not necessarily left behind, are nonetheless disadvantaged through their relative lack of agrarian resources. The possibilities of expansion-based production are smaller for households that (relatively speaking) lack land and labour, while commercial diversification also predominantly benefits male-headed households. The prospects for pro-poor agricultural growth in this sense are constrained among women. Here diversification into the non-farm sector may offer other opportunities, even if such diversification is not connected to higher agricultural productivity.

Policy Implications

Three interconnected policy implications arise from the study. First, an observation is that programmes aimed at being gender inclusive should be carefully formulated and require a clear understanding of the patterns of gender-based factors and their underlying causes. In this respect, farm and non-farm livelihood sources may vary in importance based on gender, generation, and geography.

Second, to balance farm and non-farm sources of income to enable a guided process of rural transformation requires monitoring and strategic facilitation by way of public and private investments in rural areas.

Finally, while Africa may learn from the transformation path of other continents, the path she takes will be unique and consistent with the circumstances driving change in Africa at the present and in the foreseeable future.

References

Ackah, C. (2013). 'Nonfarm employment and incomes in rural Ghana.' *Journal of International Development* 25(3): 325–39.

Adams, W. M. and R. Carter (1987). 'Small-scale irrigation in sub-Saharan Africa.' *Progress in Physical Geography* 11(1): 1–27.

AGRA (2014). 'Africa agriculture status report: climate change and smallholder agriculture in sub-Saharan Africa.' Nairobi.

Ahituv, A. and A. Kimhi (2002). 'Off-farm work and capital accumulation decisions of farmers over the life-cycle: the role of heterogeneity and state dependence.' *Journal of Development Economics* 68(2): 329–53.

Ahuja, V. (1998). 'Land degradation, agricultural productivity and common property: evidence from Côte d'Ivoire.' *Environment and Development Economics* 3(1): 7–34.

Ajani, E. N. and E. Igbokwe (2013). 'Occupational diversification among rural women in sub-saharan Africa: a review.' *African Journal of Food, Agriculture, Nutrition and Development* 13(5): 8224–37.

Alden Wily, L. (2011). '"The law is to blame": the vulnerable status of common property rights in sub-Saharan Africa.' *Development and Change* 42(3): 733–57.

Alene, A. D., V. M. Manyong, G. Omanya, H. D. Mignouna, M. Bokanga, and G. Odhiambo (2008). 'Economic efficiency and supply response of women as farm managers: comparative evidence from western Kenya.' *World Development* 36(7): 1247–60.

Alobo Loison, S. (2015). 'Rural livelihood diversification in sub-Saharan Africa: a literature review.' *The Journal of Development Studies* 51(9): 1125–38.

Alobo Loison, S. (2016). 'The dynamics of rural household livelihood diversification: panel evidence from Kenya.' *Journées de Recherches en Sciences Sociales, 10es JRSS*. Paris.

Andersson Djurfeldt, A. (2012). 'Seasonality and farm nonfarm interactions in western Kenya.' *Journal of Modern African Studies* 50(1): 1–23.

Andersson Djurfeldt, A. (2013). 'African re-agrarianization? Accumulation or pro-poor agricultural growth?' *World Development* 41: 217–31.

Andersson Djurfeldt, A. (2014). 'Multi-local livelihoods and food security in rural Africa.' *Journal of International Development* 27(4): 528–45.

Andersson Djurfeldt, A. (2015). 'Urbanization and linkages to smallholder farming in sub-Saharan Africa: Implications for Food Security.' *Global Food Security* 4(1): 1–7.

Andersson Djurfeldt, A. (2017). 'Pro-poor agricultural growth: village dynamics and commercialization pathways, a synthesis of Afrint data.' APRA Working Paper 2. I. Institute of Development Studies, Brighton.

Andersson Djurfeldt, A. and G. Djurfeldt (2013). 'Structural transformation and African smallholders: drivers of mobility within and between the farm and non-farm sectors for eight countries.' *Oxford Development Studies* 41(3): 281–306.

Andersson Djurfeldt, A., G. Djurfeldt, and J. Bergman Lodin (2013). 'Geography of gender gaps: regional patterns of income and farm–nonfarm interaction among male- and female-headed households in eight African countries.' *World Development* 48: 32–47.

Andersson Djurfeldt, A., G. Djurfeldt, and D. B. Sarpong (2014). 'Community, cohesion and context: agrarian development and religion in Eastern Region, Ghana.' *Geoforum* 52: 78–89.

Andersson Djurfeldt, A. and E. Hillbom (2016). 'Pro-poor agricultural growth: inclusion or differentiation? Village level perspectives from Zambia.' *Geoforum* 75: 220–33.

Andersson Djurfeldt, A., W. O. Mulwafu, P. Mvula, E. Hillbom, and G. Djurfeldt (2017 (re-submitted)). 'Welfare and Decision Making within matrilineal, uxorilocal land tenure systems in Malawi.'

Andersson Djurfeldt, A. and S. K. Wambugu (2011). 'In-kind transfers of maize, commercialization and household consumption in Kenya.' *Journal of Eastern African Studies* 5(3): 447–64.

Antle, J. M. (1983). 'Infrastructure and aggregate agricultural productivity: international evidence.' *Economic Development and Cultural Change* 31(3): 609–19.

Ariga, J., T. Jayne, and J. Nyoro (2006). 'Factors driving the growth in fertilizer consumption in Kenya, 1990–2005.' Tegemeo Institute Egerton University, Nairobi.

Arndt, C., K. Pauw, and J. Thurlow (2016). 'The economy-wide impacts and risks of Malawi's Farm Input Subsidy Program.' *American Journal of Agricultural Economics* 98(3): 962–80.

Arndt, C. and F. Tarp (2000). 'Agricultural technology, risk, and gender: a CGE analysis of Mozambique.' *World Development* 28(7): 1307–26.

Asfaw, S., N. McCarthy, A. Paoloantonio, R. Cavatassi, M. Amare, and L. Lipper (2015). 'Livelihood diversification and vulnerability to poverty in rural Malawi.' FAO, ESA Working Paper. Rome, FAO.

Aylward, C., P. Biscaye, K. Panhorst Harris, M. LaFayette, Z. True, L. Anderson, and T. Reynold (2015). 'Maize yield trends and agricultural policy in East Africa: macro-level literature review for six countries.' EPAR Technical Report No. 310, Evans School of Policy and Research, University of Washington.

Babatunde, R. O. and M. Qaim (2009). 'Patterns of income diversification in rural Nigeria: determinants and impacts.' *Quarterly Journal of International Agriculture* 48 (4): 305–20.

Bagachwa, M. S. D. and F. Stewart (1992). 'Rural industries and rural linkages in sub-Saharan Africa: a survey.' *Alternative Development Strategies in Sub-Saharan Africa*. F. Stewart, S. Lall, and S. Wangwe. London, Palgrave Macmillan.

Baker, C. (1962). 'Nyasaland: the history of its export trade.' *The Nyasaland Journal* 15(1): 7–35.

Bank of Tanzania (2013/14). 'Annual report.'

Barham, J. and C. Chitemi (2009). 'Collective action initiatives to improve marketing performance: lessons from farmer groups in Tanzania.' *Food policy* 34(1): 53–9.

Barrett, C. B. and M. Bezuneh (2001). 'Income diversification, poverty traps and policy shocks in Côte d'Ivoire and Kenya.' *Food Policy* 26(4): 367–84.

Barrett, C. B., T. Reardon, and P. Webb (2001). 'Nonfarm income diversification and household livelihood strategies in rural Africa: concepts, dynamics, and policy implications.' *Food policy* 26(4): 315–31.

Barrientos, S., C. Dolan, and A. Tallontire (2003). 'A gendered value chain approach to codes of conduct in African horticulture.' *World Development* 31(9): 1511–26.

Bebbington, A. (1999). 'Capitals and capabilities: a framework for analyzing peasant viability, rural livelihoods and poverty.' *World development* 27(12): 2021–44.

Beddow, J. M., T. M. Hurley, P. G. Pardey, and J. M. Alston (2014). 'Food security: yield gap. reference module in food science.' In *Encyclopedia of Agriculture and Food Systems*. N. K. Van Alfen. Oxford, Elsevier.

Bergman Lodin, J. (2012). 'Intrahousehold bargaining and distributional outcomes regarding NERICA upland rice proceeds in Hoima district, Uganda.' *Gender, Technology and Development* 16(3): 253–78.

Bernstein, H. (2010). *Class dynamics of agrarian change*. Halifax and Winnipeg Sterling, VA, Fernwood Publishing Kumarian Press.

Bernstein, H. and C. Oya (2014). 'Rural futures: how much should markets rule?' IIED Working Paper, London, IIED Natural Resources Group.

Binswanger-Mkhize, H. P., A. F. McCalla, and P. Patel (2010). 'Structural transformation and African agriculture.' *Global Journal of Emerging Market Economies* 2(2): 113–52.

Brody, A., A. Spieldoch, and G. Aboud (2014). 'Gender and food security: towards a gender-just food and nutrition security, BRIDGE report.' Brighton, Sussex, Institute of Development Studies.

Bryceson, D. (2009). 'Sub-Saharan Africa's vanishing peasantries and the specter of a global food crisis.' *Monthly Review* 61(3): 48–62.

Byerlee, D. and P. W. Heisey (1997). 'Evolution of the African maize economy.' In *Africa's emerging maize revolution*. D. Byerlee and C. K. Eicher. Colorado, Lynn Rienner Publishers.

Byerlee, D., T. S. Jayne, and R. J. Myers (2006). 'Managing food price risks and instability in a liberalizing market environment: overview and policy options.' *Food Policy* 31(4): 275–87.

Cairns, J. E., J. Hellin, K. Sonder, J. L. Araus, J. F. MacRobert, C. Thierfelder, and B. Prasanna (2013). 'Adapting maize production to climate change in sub-Saharan Africa.' *Food Security* 5(3): 345–60.

Canagarajah, S., C. Newman, and R. Bhattamishra (2001). 'Non-farm income, gender, and inequality: evidence from rural Ghana and Uganda.' *Food policy* 26(4): 405–20.

Carletto, C., D. Jolliffe, and R. Banerjee (2014). 'The emperor has no data! Agricultural statistics in sub–Saharan Africa (technical report).' Washington, DC, World Bank.

Carr, E. R. (2008). 'Men's crops and women's crops: the importance of gender to the understanding of agricultural and development outcomes in Ghana's Central Region.' *World Development* 36(5): 900–15.

Chambers, R. and G. Conway (1992). 'Sustainable rural livelihoods: practical concepts for the 21st century.' Brighton, Sussex, Institute of Development Studies.

Chant, S. (2007). *Gender, generation and poverty: exploring the feminisation of poverty thesis in Africa, Asia and Latin America*. Cheltenham, Edward Elgar.

Chayanov, A. V. (1966). *A.V. Chayanov and the theory of peasant economy*. Homewood, IL, Richard D. Irwin.

Chayanov, A. V. (1977). 'The journey of my brother Aleksei to the land of peasant utopia.' In *The Russian Peasant 1920 & 1984*. R. E. F. Smith. London, Frank Cass.

Chenery, H. and M. Syrquin (1975). *Patterns of development 1950–1970*. Oxford, Oxford University Press.

Chibwana, C., M. Fisher, and G. Shively (2012). 'Cropland allocation effects of agricultural input subsidies in Malawi.' *World Development* 40(1): 124–33.

Chikwama, C. (2004). 'Rural off-farm employment and farm investment: an analytical framework and evidence from Zimbabwe.' *African Journal of Agricultural and Resources Economics* 4(1): 1–22.

Chinsinga, B. (2007). 'Hedging food security through winter cultivation: the agronomy of dimba cultivation in Malawi.' Education Development Conference.

Chinsinga, B. and A. O'Brien (2008). 'Planting ideas: how agricultural subsidies are working in Malawi.' London, Africa Research Institute.

Chirwa, E. and A. Dorward (2013). *Agricultural input subsidies: the recent Malawi experience*. Oxford, Oxford University Press.

Chirwa, E. W. (1998). 'Fostering private food marketing and food policies after liberalization: the case of Malawi.' *World Development Studies* 12: 63.

Chirwa, E. W., I. Kumwenda, C. Jumbe, P. Chilonda, and I. Minde (2008). 'Agricultural growth and poverty reduction in Malawi: past performance and recent trends.' ReSAKSS-SA Working Paper 8.

CIMMYT (2016). 'Maize agriculture, drought, and the potential of improved seed in sub-Saharan Africa.' Blogg article. <http://dtma.cimmyt.org/index.php/about/background> accessed 30 September 2016.

CISANET (Nd). 'The pathways to crop diversification in Malawi: a case study of tobacco production.' Unpublished paper. <http://www.cisanetmw.org>.

Clark, J. T. B. S. (2013). 'Understanding the gender based productivity gap in Malawi's agricultural sector.' Master of Public Policy thesis. Washington, DC, Georgetown University.

Cochran, W. G. (2007). *Sampling techniques*. Oxford, John Wiley & Sons.

Coelli, T. and E. Fleming (2004). 'Diversification economies and specialisation efficiencies in a mixed food and coffee smallholder farming system in Papua New Guinea.' *Agricultural Economics* 31(2–3): 229–39.

Collier, P. and S. Dercon (2014). 'African agriculture in 50 years: smallholders in a rapidly changing world?' *World Development* 63: 92–101.

Corral, L. and T. Reardon (2001). 'Rural nonfarm incomes in Nicaragua.' *World Development* 29(3): 427–42.

Cragg, J. G. (1971). 'Some statistical models for limited dependent variables with application to the demand for durable goods.' *Econometrica: Journal of the Econometric Society* 39(5): 829–44.

Da Corta, L. and L. Price (2009). 'Poverty and growth in remote villages in Tanzania (2004–2008), insights from Village Voice film research.' Working Paper No. 153, Chronic Poverty Research Centre.

Davis, B., S. Di Giuseppe, and A. Zezza (2016). 'Are African households (not) leaving agriculture? Patterns of households' income sources in rural sub-Saharan Africa.' *Food Policy* 67: 153–74.

Davis, B., P. Winters, G. Carletto, K. Covarrubias, E. J. Quiñones, A. Zezza, K. Stamoulis, C. Azzarri, and S. DiGiuseppe (2010). 'A cross-country comparison of rural income generating activities.' *World Development* 38(1): 48–63.

De Janvry, A. and E. Sadoulet (2005). 'Achieving success in rural development: toward implementation of an integral approach.' *Agricultural Economics* 32(s1): 75–89.

Deere, C. D. (2010). 'Household wealth and women's poverty: conceptual and methodological issues in assessing gender inequality in asset ownership.' In *The international handbook of gender and poverty: concepts, research, policy*. S. Chant. Cheltenham, Edward Elgar.

Deere, C. D. and C. R. Doss (2006). 'The gender asset gap: what do we know and why does it matter?' *Feminist Economics* 12(1–2): 1–50.

Deere, C. D., A. D. Oduro, H. Swaminathan, and C. Doss (2013). 'Property rights and the gender distribution of wealth in Ecuador, Ghana and India.' *Journal of Economic Inequality* 11(2): 249.

Deichmann, U., Shilpi, F., and Vakis, R. (2008). 'Spatial specialization and farm-nonfarm linkages.' Policy Research Working Paper 4611. The World Bank, Washington, DC. Available at: <http://econ.worldbank.org>, last accessed 4 October 2009.

Delgado, C., P. Hazell, J. Hopkins, and V. Kelly (1994). 'Promoting intersectoral growth linkages in rural Africa through agricultural technology and policy reform.' *American Journal of Agricultural Economics* 76(5): 1166–71.

Delgado, C. L. (1998). 'Agricultural growth linkages in sub-Saharan Africa.' International Food Policy Research Institute, Research Report 107, Washington, DC, IFPRI.

Delgado, C. L. and N. W. Minot (2000). *Agriculture in Tanzania since 1986: follower or leader of growth?* Washington, DC, World Bank Publications.

Devereux, S. (2002). 'The Malawi famine of 2002.' *IDS Bulletin* 33(4): 70–8.

Diao, X., P. Hazell, and J. Thurlow (2010). 'The role of agriculture in African development.' *World Development* 38(10): 1375–83.

Djurfeldt, G., E. Aryeetey, and A. Isinika (2011). *African smallholders: food crops, markets and policy*. London, CABI.

Djurfeldt, G., E. Aryeetey, and A. Isinika (2011). 'Introduction.' In *African smallholders: food crops, markets and policy*. G. Djurfeldt, E. Aryeetey, and A. Isinika. London, CABI.

Djurfeldt, G., H. Holmén, M. Jirström, and R. Larsson, eds (2005). *The African food crisis: lessons from the Asian Green Revolution*. Wallingford, CABI Publishing.

Djurfeldt, G. and M. Jirström (2005). 'The puzzle of the policy shift: the early Green Revolution in India, Indonesia and the Philippines.' *The African food crisis: lessons from the Asian Green Revolution*. G. Djurfeldt, H. Holmen, M. Jirström, and R. Larsson. Wallingford, CABI.

Djurfeldt, G. and S. Sircar (2016). *Structural transformation and agrarian change in India*. London, Routledge.

Dorosh, P. A., S. Dradri, and S. Haggblade (2009). 'Regional trade, government policy and food security: Recent evidence from Zambia.' *Food Policy* 34(4): 350–66.

Dorward, A., E. Chirwa, V. Kelly, T. S. Jayne, R. Slater, and D. Boughton (2008). 'Evaluation of the 2006/2007 Agricultural Input Supply Programme, Malawi.' Final Report, March 2008, Imperial College, London, Wadonda Consult, Michigan State University, Overseas Development Institute, Ministry of Agriculture and Food Security, Malawi.

Dorward, A., J. Kydd, J. Morrison, and I. Urey (2004). 'A policy agenda for pro-poor agricultural growth.' *World Development* 32(1): 73–89.

Doss, C. (2013). 'Intrahousehold bargaining and resource allocation in developing countries.' *The World Bank Research Observer* 28(1): 52–78.

Doss, C. R. (2001). 'Designing agricultural technology for African women farmers: lessons from 25 years of experience.' *World Development* 29(12): 2075–92.

Doss, C. R. (2002). 'Men's crops? Women's crops? The gender patterns of cropping in Ghana.' *World Development* 30(11): 1987–2000.

Doss, C. R. (2014). 'Data needs for gender analysis in agriculture.' In *Gender in agriculture*. A. Quisumbing, R. Meinzen-Dick, T. Raney, A. Croppenstedt, J. Behrman, and A. Peterman. Dordrecht, Springer.

Doss, C. R. and M. Morris (2001). 'How does gender affect the adoption of agricultural innovations? The case of improved maize technology in Ghana.' *Agricultural Economics* 25(1): 27–39.

Dzanku, F. M. (2015). 'Transient rural livelihoods and poverty in Ghana.' *Journal of Rural Studies* 40: 102–10.

Dzanku, F. M. and P. Aidam (2013). 'Agricultural sector development: policies and options.' In *Policies and Options for Ghana's Economic Development*. K. Ewusi. Accra, ISSER, University of Ghana at Legon.

Dzanku, F. M., M. Jirström, and H. Marstorp (2015). 'Yield gap-based poverty gaps in rural Sub-Saharan Africa.' *World Development* 67: 336–62.

Dzanku, F. M. and B. D. Sarpong (2014). 'Household-level farm–nonfarm linkages and household welfare implications. Afrint III: Ghana micro report.' Available at: <http://www.keg.lu.se/en/research/research-projects/current-research-projects/afrint>, Afrint Project.

Dzanku, F. M. and C. Udry (2017). 'Flickering decades of agriculture and agricultural policy.' In *The Economy of Ghana Sixty Years After Independence*. Oxford, Oxford University Press.

ECA (2016). 'MDGs to agenda 2063/SDGs. Transition report. Towards an integrated and coherent approach to sustainable development in Africa.' Addis Ababa, Economic Commission for Africa, African Development Bank, African Union Commission UNDP's Regional Bureau for Africa.

Elbers, C. and P. Lanjouw (2001). 'Intersectoral transfer, growth, and inequality in rural Ecuador.' *World Development* 29(3): 481–96.

Ellis, F. (2007). 'Strategic dimensions of rural poverty reduction in sub-Saharan Africa.' Paper presented at the workshop Rural Development Retrospect and Prospect: A Workshop for Judith Heyer. Oxford.

Ellis, F. (2009). 'Strategic dimensions of rural poverty reduction in sub-Saharan Africa.' In *The Comparative Political Economy of Development: Africa and South Asia*. B. Harriss-White and J. Heyer. London, Routledge.

Ellis, F. and H. A. Freeman (2004). 'Rural livelihoods and poverty reduction strategies in four African countries.' *Journal of Development Studies* 40(4): 1–30.

Ellis, F. and N. Mdoe (2003). 'Livelihoods and rural poverty reduction in Tanzania.' *World Development* 31(8): 1367–84.

Eriksson, G. (1993). 'Peasant response to price incentives in Tanzania: A theoretical and empirical investigation.' Research Report No. 91. Uppsala, Scandinavian Institute of African Studies.

Evans, H. E. and P. Ngau (1991). 'Rural–urban relations, household income diversification and agricultural productivity.' *Development and Change* 22(3): 519–45.

FAO (2010). 'FAO gender and land rights database.' Rome, Food and Agriculture Organization.

FAO (2011). 'Women in agriculture: closing the gender gap for development.' Rome, Food and Agriculture Organization.

FAO (2014). *The state of food insecurity in the world 2014: strengthening the enabling environment for food security and nutrition*. Rome, Food and Agriculture Organization.

FAO (2016). 'FAOstat data.' Rome, Food and Agriculture Organization.

FAO and DWFI (2015). 'Yield gap analysis of field crops: methods and case studies.' FAO Water Reports, Rome, Food and Agriculture Organization.

Finan, F., E. Sadoulet, and A. De Janvry (2005). 'Measuring the poverty reduction potential of land in rural Mexico.' *Journal of Development Economics* 77(1): 27–51.

Fischer, E. and M. Qaim (2012). 'Gender, agricultural commercialization, and collective action in Kenya.' *Food Security* 4(3): 441–53.

Frelat, R., S. Lopez-Ridaura, K. E. Giller, M. Herrero, S. Douxchamps, A. A. Djurfeldt, O. Erenstein, B. Henderson, M. Kassie, and B. K. Paul (2016). 'Drivers of household food availability in sub-Saharan Africa based on big data from small farms.' *Proceedings of the National Academy of Sciences* 113(2): 458–63.

Fuglie, K. O. (2011). 'Agricultural Productivity in sub-Saharan Africa.' In *The food and financial crises in sub-Saharan Africa: origins, impacts and policy implications*. M. B. Ndulo. Wallingford, CABI.

Gann, L. H. (1969). *A history of Northern Rhodesia: early days to 1953*, Humanities Press.

Ghosh, J. (2010). 'The unnatural coupling: food and global finance.' *Journal of Agrarian Change* 10(1): 72–86.

Gibson, J. and S. Olivia (2010). 'The effect of infrastructure access and quality on non-farm enterprises in rural Indonesia.' *World Development* 38(5): 717–26.

Goldstein, M. and C. Udry (2008). 'The profits of power: land rights and agricultural investments in Ghana.' *Journal of Political Economy* 116(6): 981–1022.

Gollin, D., R. Jedwab, and D. Vollrath (2016). 'Urbanization with and without industrialization.' *Journal of Economic Growth* 21(1): 35–70.

Gollin, D. and R. Rogerson (2014). 'Productivity, transport costs and subsistence agriculture.' *Journal of Development Economics* 107: 38–48.

Gotschi, E., J. Njuki, and R. Delve (2009). 'Equal numbers, equal chances? A case study of gender differences in the distribution of social capital in smallholder farmer groups in Búzi District, Mozambique.' *European Journal of Development Research* 21(2): 264–82.

Govereh, J. and T. S. Jayne (2003). 'Cash cropping and food crop productivity: synergies or trade-offs?' *Agricultural Economics* 28(1): 39–50.

Government of Malawi (2007). 'Agricultural Development Programme.' Lilongwe, Ministry of Agriculture and Food Security.

Government of Malawi (2012). 'Brief IRLAD, no. 96390.' Lilongwe.

Government of Malawi (2015). 'Irrigation, Rural Livelihoods and Agricultural Development Project.' IRLAD Implementation Completion Report.

Guendel, S. (2009). 'What are "women's crops", and why?' FAO Gender Insight, blog.

Haggblade, S. (2007). 'Alternative perceptions of the rural nonfarm economy.' *Transforming the rural nonfarm economy: opportunities and threats in the developing world.* S. Haggblade, P. Hazell, and T. Reardon. Baltimore, Johns Hopkins University Press.

Haggblade, S., P. Hazell, and J. Brown (1989). 'Farm-nonfarm linkages in rural sub-Saharan Africa.' *World Development* 17(8): 1173–201.

Haggblade, S., P. Hazell, and T. Reardon (2007). *Transforming the rural nonfarm economy: opportunities and threats in the developing world.* Baltimore, Johns Hopkins University Press.

Haggblade, S., P. Hazell, and T. Reardon (2010). 'The rural non-farm economy: prospects for growth and poverty reduction.' *World Development* 38(10): 1429–41.

Handschuch, C. and M. Wollni (2013). 'Traditional food crop marketing in sub-Saharan Africa: does gender matter?' Poverty, Equity and Growth-Discussion Papers, Courant Research Centre.

Harrigan, J. (2001). *From dictatorship to democracy: economic policy in Malawi 1964–2000.* Farnham, Ashgate.

Harrigan, J. (2003). 'U-turns and full circles: two decades of agricultural reform in Malawi 1981–2000.' *World Development* 31(5): 847–63.

Harrigan, J. (2005). 'Food security policies and starter pack: a challenge for donors.' In *Starter packs: a strategy to fight hunger in developing countries.* S. Levy. Wallingford, CABI.

Harris, J. M., S. C. Blank, K. Erickson, and C. Hallahan (2010). *Off-farm income and investments in farm assets: a double-hurdle approach.* Selected paper prepared for presentation at the AAEA, CAES, and WAEA Joint Annual Meeting, Denver, Colorado.

Harvest Choice. (2011). 'Total livestock population (TLU) (2005).' Available at: <http://harvestchoice.org/node/4788>.

Havnevik, K., D. Bryceson, L. E. Birgegård, P. Matondi, and A. Beyene (2007). 'African agriculture and the World Bank: development or impoverishment?' Policy Dialogue No. 1. Uppsala, Nordic Africa Institute.

Hayami, Y. and V. W. Ruttan (1971). *Agricultural development: an international perspective.* Baltimore and London, Johns Hopkins University Press.

Hazell, P. (2013). 'Comparative study of trends in urbanization and changes in farm size in Africa and Asia: implications for agricultural research.' Background paper for the ISPC Foresight Study on Farm Size and Urbanization.

Hazell, P., S. Haggblade, and T. Reardon (2007). 'Structural transformation of the rural non farm economy.' In *Transforming the Rural Nonfarm Economy: Opportunities and*

Threats in the Developing World. S. Haggblade, P. Hazell, and T. Reardon. Baltimore, Johns Hopkins University Press.

Hazell, P. and B. Hojjati (1995). 'Farm/non-farm growth linkages in Zambia.' *Journal of African Economies* 4(3): 406–35.

Hazell, P. and C. Poulton (2007). 'All-Africa review of experiences with commercial agriculture: case study on food staples.' Background paper for the Competitive Commercial Agriculture in Sub-Saharan Africa (CCAA) Study. Washington, DC, World Bank.

Hazell, P. and A. Rahman (2014). *New directions for smallholder agriculture*. Oxford, Oxford University Press.

Helbling, T., V. Mercer-Blackman, and K. Cheng (2008). 'Commodity boom: riding a wave.' Washington, DC, Finance and Development, IMF.

Hengsdijk, H., A. Franke, M. van Wijk, and K. E. Giller (2014). 'How small is beautiful? Food self-sufficiency and land gap analysis of smallholders in humid and semi-arid sub-Saharan Africa.' Plant Research International, Business Unit Agrosystems Research.

Hertz, T. (2009). 'The effect of nonfarm income on investment in Bulgarian family farming.' *Agricultural Economics* 40(2): 161–76.

Hill, C. (2011). 'Enabling rural women's economic empowerment: institutions, opportunities, and participation.' Background paper prepared for the Expert Group Meeting on Enabling Rural Women's Economic Empowerment: Institutions, Opportunities and Participation, Accra, UN Women in Cooperation with FAO, IFAD, and WFP.

Hill, R. V. and M. Vigneri (2014). 'Mainstreaming gender sensitivity in cash crop market supply chains.' In *Gender in agriculture*. A. Quisumbing, R. Meinzen-Dick, T. Raney, A. Croppenstedt, J. Behrman, and A. Peterman. Dordrecht, Springer.

Holden, S. T. and R. W. Lunduka (2013). 'Who benefits from Malawi's targeted farm input subsidy program?' *Forum for Development Studies* 40(1): 1–25.

Holmén, H. (2005). 'Spurts in production: Africa's limping green revolution.' In *African food crisis: the relevance of the Asian Green Revolution*. G. Djurfeldt, H. Holmén, M. Jirström, and R. Larsson. Wallingford, CABI.

Hope, R., P. Dixon, and G. Von Maltitz (2003). 'The role of improved domestic water supply in livelihoods and poverty reduction in Limpopo Province, South Africa.' Proceedings of International Symposium on Water, Poverty and Productive Uses of Water at the Household Level.

Horrell, S. and P. Krishnan (2009). 'Poverty and productivity in female headed households in Zimbabwe.' *Journal of Development Studies* 43(8): 1351–80.

IFAD (1998). 'Agricultural implements used by women farmers in Africa.' Rome, IFAD.

Isinika, A., G. Ashimogo, and J. E. D. Mlwangwa (2004). 'Market conditions, infrastructure and agricultural intensification in Tanzania since independence.' Paper prepared for The African Food Crisis: The Relevance of Asian Models, workshop, Nairobi.

Isinika, A. and E. E. Msuya (2016). 'Gender and inclusion: the dynamics of nonfarm–farm linkages for pro-poor agricultural growth, Tanzania.' Afrint III report.

Isinika, A. C., G. C. Ashimogo, and J. E. Mlangwa (2005). 'From Ujamaa to structural adjustment: agricultural intensification in Tanzania.' In *The African food crisis: lessons from the Asian Green Revolution*. G. Djurfeldt, H. Holmén, M. Jirström, and R. Larsson. London, CABI.

Isinika, A. C., E. E. Msuya, G. Djurfeldt, and E. Aryeetey (2011). 'Addressing food self-sufficiency in Tanzania: a balancing act of policy coordination.' *African smallholders: food crops, markets and policy*. E. Aryeetey, G. Djurfeldt, and A. Isinika. London, CABI.

Jackson, C. (1999). 'Men's work, masculinities and gender divisions of labour.' *Journal of Development Studies* 36(1): 89–108.

Jackson, C. (2003). 'Gender analysis of land: beyond land rights for women?' *Journal of Agrarian Change* 3(4): 453–80.

Jackson, C. (2007). 'Resolving risk? Marriage and creative conjugality.' *Development and Change* 38(1): 107–29.

Jagger, P. and J. Pender (2006). 'Influences of programs and organizations on the adoption of sustainable land management technologies in Uganda.' In *Strategies for sustainable land management in the East African highlands*. J. Pender, F. Place, and S. Ehui. Washington, DC, International Food Policy Research Institute.

Jayne, T., J. Chamberlin, and D. D. Headey (2014). 'Land pressures, the evolution of farming systems, and development strategies in Africa: A synthesis.' *Food Policy* 48: 1–17.

Jayne, T., S. Haggblade, N. Minot, and S. Rashid (2011). 'Agricultural commercialization, rural transformation and poverty reduction: what have we learned about how to achieve this?' Synthesis report prepared for the African Agricultural Markets Programme Policy Symposium, Alliance for Commodity Trade in Eastern and Southern Africa, April.

Jayne, T. S. and S. Jones (1997). 'Food marketing and pricing policy in Eastern and Southern Africa: a survey.' *World Development* 25(9): 1505–27.

Jayne, T. S., D. Mather, and E. Mghenyi (2006a). 'Smallholder farming under increasingly difficult circumstances: policy and public investment priorities for Africa.' MSU International Development Working Paper. East Lansing, Department of Agricultural Economics, Michigan State University.

Jayne, T. S., D. Mather, and E. Mghenyi (2010). 'Principal challenges confronting smallholder agriculture in sub-Saharan Africa.' *World Development* 38(10): 1384–98.

Jayne, T. S., B. Zulu, and J. J. Nijhoff (2006b). 'Stabilizing food markets in Eastern and Southern Africa.' *Food Policy* 31: 328–41.

Jirström, M., A. Andersson, and G. Djurfeldt (2011). 'Smallholders caught in poverty: flickering signs of agricultural dynamism.' *African smallholders: food crops, markets and policy*. E. Aryeetey, G. Djurfeldt, and A. Isinika. London, CABI.

Jirström, M., F.-M. Rundquist, S. Nasrin, and B. Holmquist (2012). 'Changes in diversification profiles and its impacts on total income among rice farming households in the Mekong and Red River Deltas in Vietnam 1999–2005.' The Association of American Geographers annual meeting, 2012.

Johnson, N. L., C. Kovarik, R. Meinzen-Dick, J. Njuki, and A. Quisumbing (2016). 'Gender, assets, and agricultural development: lessons from eight projects.' *World Development* 83: 295–311.

Kasente, D., M. Lockwood, J. Vivian, and A. Whitehead (2002). 'Gender and the expansion of non-traditional agricultural exports in Uganda.' In *Shifting burdens:*

gender and agrarian change under neoliberalism. S. Razavi. Bloomfield, CT, Kumarian Press.

Katega, I. and C. Lifuliro (2014). 'Rural non-farm activities and poverty alleviation in Tanzania: a case study of two villages in Chamwino and Bahi districts of Dodoma region.' REPOA Working Paper 14/4.

Kerr, R. B. (2005). 'Food security in northern Malawi: gender, kinship relations and entitlements in historical context.' *Journal of Southern African Studies* 31(1): 53–74.

Kherallah, M. and K. Govindan (1999). 'The sequencing of agricultural market reforms in Malawi.' *Journal of African Economies* 8(2): 125–51.

Kibaara, B., J. Ariga, J. Olwande, and T. S. Jayne (2009). 'Trends in Kenyan agricultural productivity: 1997–2007.' Food Security Collaborative Working Paper No. 56117, Michigan State University.

Kijima, Y., T. Matsumoto, and T. Yamano (2006). 'Nonfarm employment, agricultural shocks, and poverty dynamics: evidence from rural Uganda.' *Agricultural Economics* 35(s3): 459–67.

Kilic, T., C. Carletto, J. Miluka, and S. Savastano (2009). 'Rural nonfarm income and its impact on agriculture: evidence from Albania.' *Agricultural Economics* 40(2): 139–60.

Kilic, T., A. Palacios-Lopez, and M. Goldstein (2015). 'Caught in a productivity trap: a distributional perspective on gender differences in Malawian agriculture.' *World Development* 70: 416–63.

Kimenju, S. C. and D. Tschirley (2009). 'Agriculture and livelihood diversification in Kenyan rural households.' Tegemeo Institute of Agricultural Policy and Development, Nairobi.

Kinkingninhoun-Mêdagbé, F. M., A. Diagne, F. Simtowe, A. R. Agboh-Noameshie, and P. Y. Adégbola (2010). 'Gender discrimination and its impact on income, productivity, and technical efficiency: evidence from Benin.' *Agriculture and Human Values* 27(1): 57–69.

Kish, L. (1965). *Survey sampling*. New York, John Wiley.

Kish, L. (1987). *Statistical design for research*. New York, John Wiley.

Kishindo, P. (2004). 'Customary land tenure and the new land policy in Malawi.' *Journal of Contemporary African Studies* 22(2): 213–25.

Kolavalli, S., K. Flaherty, R. Al-Hassan, and K. O. Baah (2010). 'Do Comprehensive Africa Agriculture Development Program (CAADP) processes make a difference to country commitments to develop agriculture? The case of Ghana.' Discussion Paper 1006, International Food Policy Research Institute.

Kristjanson, P., A. Waters-Bayer, N. Johnson, A. Tipilda, J. Njuki, I. Baltenweck, D. Grace, and S. Macmillan (2014). 'Livestock and women's livelihoods.' In *Gender in agriculture*. A. Quisumbing, R. Meinzen-Dick, T. Raney, A. Croppenstedt, J. Behrman, and A. Peterman. Dordrecht, Springer.

Kydd, J. (1984). 'Malawi in the 1970s: development policies and economic change.' Paper for Malawi: An Alternative Pattern of Development, conference, Centre for African Studies, University of Edinburgh.

Lamb, R. L. (2003). 'Fertilizer use, risk, and off-farm labor markets in the semi-arid tropics of India.' *American Journal of Agricultural Economics* 85(2): 359–71.

Lanjouw, P. and G. Feder (2001). 'Rural non-farm activities and rural development: from experience towards strategy.' Working Paper 22818. Washington, DC, World Bank.

Larsson, R. (2001). 'Between crisis and opportunity: livelihoods, diversification and inequality among the Meru of Tanzania.' Lund Dissertations in Sociology, 41.

Larsson, R. (2005). 'Crisis and potential in smallholder food production: evidence from micro level.' In *The African food crisis: lessons from the Asian Green Revolution*. G. Djurfeldt, H. Holmén, M. Jirström, and R. Larsson. London, CABI.

Lastarria-Cornhiel, S., J. A. Behrman, R. Meinzen-Dick, and A. R. Quisumbing (2014). 'Gender equity and land: toward secure and effective access for rural women.' In *Gender in agriculture*. A. Quisumbing, R. Meinzen-Dick, T. Raney, A. Croppenstedt, J. Behrman, and A. Peterman. Dordrecht, Springer.

Lay, J. and T. O. Mahmoud (2008). 'Few opportunities, much desperation: the dichotomy of non-agricultural activities and inequality in western Kenya.' *World Development* 36(12): 2713–32.

Lele, U. (1990). 'Structural adjustment, agricultural development and the poor: some lessons from the Malawian experience.' *World Development* 18(9): 1207–19.

Lenin, V. (1899, 1960). *The development of capitalism in Russia*. Moscow, Foreign Languages Publishing House.

Levy, S. (2005). *Starter packs: a strategy to fight hunger in developing countries? Lessons from the Malawi experience 1998–2003*. Wallingford, CABI.

Levy, S. and C. Barahona (2002). 'Briefing notes starter pack and TIP: what does evidence tell us.' Available at: <http://www.reading.ac.uk/~snsbarah/TIP2>.

Lewis, A. W. (1954). 'Economic development with unlimited supplies of labour.' *The Manchester School* 22(2): 139–91.

Lilja, N., J. H. Sanders, C. A. Durham, H. De Groote, and I. Dembélé (1996). 'Factors influencing the payments to women in Malian agriculture.' *American Journal of Agricultural Economics* 78(5): 1340–5.

Lipton, M. (2005). 'The family farm in a globalizing world: the role of crop science in alleviating poverty.' Washington, DC, IFPRI.

London, L., S. De Grosbois, C. Wesseling, S. Kisting, H. A. Rother, and D. Mergler (2002). 'Pesticide usage and health consequences for women in developing countries: out of sight out of mind?' *International Journal of Occupational and Environmental Health* 8(1): 46–59.

López, R. (1998). 'The tragedy of the commons in Côte d'Ivoire agriculture: empirical evidence and implications for evaluating trade policies.' *World Bank Economic Review* 12: 105–32.

Losch, B. (2012). 'Agriculture: the key to the employment challenge.' Perspective Development, CIRAD, Montpellier, CIRAD.

Losch, B., S. Fréguin-Gresh, and E. T. White (2012). 'Structural transformation and rural change revisited: challenges for late developing countries in a globalizing world.' World Bank.

Losch, B., G. Magrin, and J. Imbernon (2013). *A new emerging rural world: an overview of rural change in Africa*. Atlas for the NEPAD Rural Futures Programme. South Africa, NEPAD.

Maertens, M. and J. F. Swinnen (2009). 'Trade, standards, and poverty: evidence from Senegal.' *World Development* 37(1): 161–78.

Marenya, P. P., W. Kosura-Olouch, et al. (2003). 'Education, nonfarm income and farm investment in land scarce Western Kenya.' Basis Brief, Madison.

Mason, N. M., T. S. Jayne, and R. J. Myers (2012). 'Smallholder behavioral responses to marketing board activities in a dual channel marketing system: the case of maize in Zambia.' Selected Paper prepared for Presentation at the International Association of Agricultural Economists Triennial Conference, Foz do Iguaçu, Brazil.

Mason, N. M. and R. Myers, J. (2013). 'The effects of the food reserve agency on maize market prices in Zambia.' *Agricultural Economics* 44(2): 203–16.

Masters, W. A. (2011). 'Africa's turnaround: from crisis to opportunity in African agriculture.' In *Food and Financial Crises in Sub-Saharan Africa*. D. R. D. Lee and M. Ndulo. Wallingford, CABI.

Masters, W. A., A. Andersson Djurfeldt, C. De Haan, P. Hazell, T. Jayne, M. Jirström, and T. Reardon (2013). 'Urbanization and farm size in Asia and Africa: implications for food security and agricultural research.' *Global Food Security* 2(3): 156–65.

Mathenge, M. K., M. Smale, and D. Tschirley (2015). 'Off-farm employment and input intensification among smallholder maize farmers in Kenya.' *Journal of Agricultural Economics* 66(2): 519–36.

Matsumoto, T., Y. Kijima, and T. Yamano (2006). 'The role of local nonfarm activities and migration in reducing poverty: evidence from Ethiopia, Kenya, and Uganda.' *Agricultural Economics* 35(s3): 449–58.

McCann, J. C. (2005). *Maize and grace: Africa's encounter with a New World Crop 1500–2000*. Cambridge, MA, Harvard University Press.

McCracken, J. (1984). 'Share-cropping in Malawi: the visiting tenant system in the Central Province, c. 1920–1968.' Paper for Malawi: An Alternative Pattern of Development, conference, Centre of African Studies, University of Edinburgh.

McCullough, E. B. (2016). 'Labor productivity and employment gaps in sub-Saharan Africa.' *Food Policy* 67: 133–52.

McGranahan, G., D. Mitlin, D. Satterthwaite, C. Tacoli, and I. Turok (2009). 'Africa's urban transition and the role of regional collaboration.' Human Settlement's Working Paper Series. Theme: Urban Change-5. Cape Town and London, African Centre for Cities and International Institute for Environment and Development.

McMillan, M. and K. Harttgen (2014). 'What is driving the "African growth miracle"?' Working Paper Series No 209. Tunis, African Development Bank.

Meinzen-Dick, R., N. Johnson, A. R. Quisumbing, J. Njuki, J. A. Behrman, D. Rubin, A. Peterman, and E. Waithanji (2014). 'The gender asset gap and its implications for agricultural and rural development.' In *Gender in agriculture*. A. Quisumbing, R. Meinzen-Dick, T. Raney, A. Croppenstedt, J. Behrman, and A. Peterman. Dordrecht, Springer.

Meinzen-Dick, R. and E. Mwangi (2009). 'Cutting the web of interests: pitfalls of formalizing property rights.' *Land Use Policy* 26(1): 36–43.

Mellor, J. W. (1995). *Agriculture on the road to industrialization*. Baltimore, Johns Hopkins University Press.

Mellor, J. W. (2014). 'High rural population density Africa: what are the growth requirements and who participates?' *Food Policy* 48: 66–75.

Mitchell, D. (2008). 'A note on rising food prices.' World Bank Development Prospects Group. Washington, DC, World Bank.

Mkandawire, M. L. and S. Trust (1999). *Poverty and macro economic management in Malawi*. Harare, SAPES Books.

Moore, H. L. and M. Vaughan (1994). *Cutting down trees: gender, nutrition, and agricultural change in the northern province of Zambia, 1890–1990*. London, James Currey.

Msuya, E. E. and A. Isinika (2016). 'Micro study, Tanzania.' Afrint III Report.

Mueller, N. D. and S. Binder (2015). 'Closing yield gaps: consequences for the global food supply, environmental quality and food security.' *Daedalus* 144(4): 45–56.

Muyanga, M. and T. Jayne (2014). 'Is small still beautiful? The farm size–productivity relationship revisited in Kenya.' World Bank conference on Land Policy in Africa, Washington, DC.

Mwase, W., C. Jumbe, F. Gasc, T. Owiyo, D. Manduwa, J. Nyaika, K. Kwapata, and B. Maonga (2014). 'Assessment of agricultural sector policies and climate change in Malawi: the nexus between climate change related policies, research and practice.' *Journal of Sustainable Development* 7(6): 195.

Nasrin, S., J. B. Lodin, M. Jirström, B. Holmquist, A. A. Djurfeldt, and G. Djurfeldt (2015). 'Drivers of rice production: evidence from five sub-Saharan African countries.' *Agriculture & Food Security* 4(1): 12.

Negin, J., R. Remans, S. Karuti, and J. C. Fanzo (2009). 'Integrating a broader notion of food security and gender empowerment into the African Green Revolution.' *Food Security* 1(3): 351–60.

Nelson, A. (2008). 'Travel time to major cities: a global map of accessibility.' Ispra, Italy, Global Environment Monitoring Unit-Joint Research Centre of the European Commission.

NEPAD (2013). 'Implementing the Comprehensive Africa Agriculture Development Programme and restoring food security in Africa: the roadmap.' South Africa, NEPAD.

Neumann, K., P. H. Verburg, E. Stehfest, and C. Müller (2010). 'The yield gap of global grain production: a spatial analysis.' *Agricultural Systems* 103(5): 316–26.

Newman, C. and S. Canagarajah (2000). 'Non-farm employment, poverty, and gender linkages: evidence from Ghana and Uganda.' Working Draft. Washington, DC, World Bank.

Ng'ong'ola, C. (1986). 'Malawi's agricultural economy and the evolution of legislation on the production and marketing of peasant economic crops.' *Journal of Southern African Studies* 12(2): 240–62.

O'Laughlin, B. (2007). 'A bigger piece of a very small pie: intrahousehold resource allocation and poverty reduction in Africa.' *Development and Change* 38(1): 21–44.

Ochieng, J., B. Knerr, G. Owuor, and E. Ouma (2016). 'Commercialisation of food crops and farm productivity: evidence from smallholders in Central Africa.' *Agrekon* 55(4): 458–82.

OECD/FAO (2016). 'Agriculture in sub-Saharan Africa: prospects and challenges for the next decade.' In *OECD-FAO Agricultural Outlook 2016–2025*. Paris, OECD Publishing.

Ogada, M. J. (2010). 'Production risk and farm technology in rain-fed maize production in semi-arid lands of Kenya.' Discussion Paper Series, DP/120/2010. Kenya Institute for Public Policy Research and Analysis (KIPPRA), Nairobi.

Ogada, M. J., W. Nyangena, and M. Yesuf (2010). 'Production risk and farm technology adoption in the rain-fed, semi-arid lands of Kenya.' *AfJARE* 4(1): 154–74.

Okali, C. (2012). 'Gender analysis: engaging with rural development and agricultural policy processes.' Future Agricultures Working Paper 026.

Omamo, S. W. (1998). 'Transport costs and smallholder cropping choices: an application to Siaya District, Kenya.' *American Journal of Agricultural Economics* 80(1): 116–23.

Ombakah, E. P. J. (2014). 'Roles of women's non-farm income generating activities to household food security: Case of Bagamoyo District.' Sokoine University of Agriculture.

Orr, A. and B. Mwale (2001). 'Adapting to adjustment: smallholder livelihood strategies in southern Malawi.' *World Development* 29(8): 1325–43.

Oseni, G., P. Corral, M. Goldstein, and P. Winters (2015). 'Explaining gender differentials in agricultural production in Nigeria.' *Agricultural Economics* 46(3): 285–310.

Oseni, G. and P. Winters (2009). 'Rural nonfarm activities and agricultural crop production in Nigeria.' *Agricultural Economics* 40(2): 189–201.

Palacios-López, A., L. Christiaensen, and T. Kilic (2016). 'How much of the labor in African agriculture is provided by women?' *Food Policy* 67: 52–63.

Palacios-López, A. and R. López (2015). 'The gender gap in agricultural productivity: the role of market imperfections.' *Journal of Development Studies* 51(9): 1175–92.

Patel, S. (2011). 'Tanzania National Agricultural Input Voucher Scheme (NAIVS) impact evaluation, baseline report.' Washington, DC, World Bank.

Peskett, L., R. Slater, C. Stevens, and A. Dufey (2007). 'Biofuels, agriculture and poverty reduction: natural resource perspectives.' London, Overseas Development Institute.

Peterman, A., J. A. Behrman, and A. R. Quisumbing (2014). 'A review of empirical evidence on gender differences in nonland agricultural inputs, technology, and services in developing countries.' In *Gender in agriculture*. A. Quisumbing, R. Meinzen-Dick, T. Raney, A. Croppenstedt, J. Behrman, and A. Peterman. Dordrecht, Springer.

Peterman, A., A. Quisumbing, J. Behrman, and E. Nkonya (2011). 'Understanding the complexities surrounding gender differences in agricultural productivity in Nigeria and Uganda.' *Journal of Development Studies* 47(10): 1482–509.

Peterman, A., A. R. Quisumbing, and J. Behrman (2010). 'Review of empirical evidence on gender differences in non-land agricultural inputs, technology and services in developing countries.' Background paper prepared for The State of Food and Agriculture 2010–2011. Rome, FAO.

Peters, P. E. (1997). 'Against the Odds Matriliny, land and gender in the Shire Highlands of Malawi.' *Critique of Anthropology* 17(2): 189–210.

Peters, P. E. (2004a). 'Inequality and social conflict over land in Africa.' *Journal of Agrarian Change* 4(3): 269–314.

Peters, P. E. (2004b). 'Informal irrigation in Lake Chilwa basin: streambank and wetland gardens.' BASIS Report. Madison, WI, BASIS CRSP.

Peters, P. E. (2010). '"Our daughters inherit our land, but our sons use their wives' fields": matrilineal-matrilocal land tenure and the New Land Policy in Malawi.' *Journal of Eastern African Studies* 4(1): 179–99.

Peters, P. E. (2013). 'Land appropriation, surplus people and a battle over visions of agrarian futures in Africa.' *Journal of Peasant Studies* 40(3): 537–62.

Peters, P. E. and D. Kambewa (2007). 'Whose security? Deepening social conflict over "customary" land in the shadow of land tenure reform in Malawi.' *Journal of Modern African Studies* 45(03): 447–72.

Pfeiffer, L., A.López-Feldman, and J. E. Taylor (2009). 'Is off-farm income reforming the farm? Evidence from Mexico.' *Agricultural Economics* 40(2): 125–38.

Piesse, J. and C. Thirtle (2009). 'Three bubbles and a panic: an explanatory review of recent food commodity price events.' *Food policy* 34(2): 119–29.

Pinstrup-Andersen, P., R. Pandya-Lorch, and M. W. Rosegrant (1997). 'The world food situation: recent developments, emerging issues, and long-term prospects.' Washington, DC, IFPRI.

Platteau, J.-P., Y. Hayami, and P. Dasgupta (1998). 'Resource endowments and agricultural development: Africa versus Asia.' In *The institutional foundations of East Asian economic development*. Y. Hayami and M. Aoki. London, Palgrave Macmillan.

Poulton, C. (draft 2017). 'What is agricultural commercialisation, why is it important and how can we measure it?' Background paper prepared for the DFID-funded Agricultural Policy Research in Africa Consortium. London, School of Oriental and African Studies.

Quisumbing, A. R. (2010). 'Gender and household decision-making in developing countries: a review of evidence.' In *The international handbook of gender and poverty*. S. Chant. Cheltenham, Edward Elgar.

Quisumbing, A. R. and J. A. Maluccio (2003). 'Resources at marriage and intrahousehold allocation: evidence from Bangladesh, Ethiopia, Indonesia, and South Africa.' *Oxford Bulletin of Economics and Statistics* 65(3): 283–327.

Quisumbing, A. R., R. Meinzen-Dick, T. L. Raney, A. Croppenstedt, J. A. Behrman, and A. Peterman (2014). 'Closing the knowledge gap on gender in agriculture.' In *Gender in agriculture*. A. Quisumbing, R. Meinzen-Dick, T. Raney, A. Croppenstedt, J. Behrman, and A. Peterman. Dordrecht, Springer.

Quisumbing, A. R. and L. Pandolfelli (2010). 'Promising approaches to address the needs of poor female farmers: resources, constraints and interventions.' *World Development* 38(4): 581–92.

Quisumbing, A. R., D. Rubin, C. Manfre, E. Waithanji, M. van den Bold, D. Olney, N. Johnson, and R. Meinzen-Dick (2015). 'Gender, assets, and market-oriented agriculture: learning from high-value crop and livestock projects in Africa and Asia.' *Agriculture and Human Values* 32(4): 705–25.

Ragasa, C., A. Chapato, and S. Kolavalli (2014). 'Maize productivity in Ghana.' Ghana Strategy Support Program. Policy Note 5. Washington, DC, IFPRI.

Rakodi, C. (1999). 'A capital assets framework for analysing household livelihood strategies: implications for policy.' *Development Policy Review* 17(3): 315–42.

Rasbash, J., W. Browne, M. Healy, B. Cameron, and C. Charlton (2010). 'MLWiN (Version 2.22).' Bristol, Centre for Multilevel Modelling, University of Bristol.

RATES (2003). 'Maize market assessment and baseline study for Malawi.' RATES, Nairobi.

Ravallion, M. and S. Chen (2004). 'Learning from success: understanding China's (uneven) progress against poverty.' *Finance & Development* 41(14).

Ravallion, M. and G. Datt (2002). 'Why has economic growth been more pro-poor in some states of India than others?' *Journal of Development Economics* 68: 381–400.

Razavi, S. (2003). 'Introduction: agrarian change, gender and land rights.' *Journal of Agrarian Change* 3(1–2): 2–32.

Razavi, S. (2009). 'Engendering the political economy of agrarian change.' *Journal of Peasant Studies* 36(1): 197–226.

Reardon, T. (1997). 'Using evidence of household income diversification to inform study of the rural nonfarm labor market in Africa.' *World Development* 25(5): 735–47.

Reardon, T., C. B. Barrett, J. A. Berdegué, and J. F. M. Swinnen (2009). 'Agrifood industry transformation and small farmers in developing countries.' *World Development* 37: 1717–27.

Reardon, T., J. Berdegué, C. B. Barrett, and K. Stamoulis (2007). 'Household income diversification into rural nonfarm activities.' In *Transforming the rural nonfarm economy: opportunities and threats in the developing world.* S. Haggblade, P. Hazell, and T. Reardon. Baltimore, Johns Hopkins University Press.

Reardon, T., E. Crawford, and V. Kelly (1994). 'Links between nonfarm income and farm investment in African households: adding the capital market perspective.' *American Journal of Agricultural Economics* 76(5): 1172–6.

Renkow, M. (2007). 'Cities, towns, and the rural nonfarm economy.' In *Transforming the rural nonfarm economy: opportunities and threats in the developing world.* S. Haggblade, P. Hazell, and T. Reardon. Baltimore, Johns Hopkins University Press.

Republic of Kenya (2004). 'Strategy to revitalize agriculture.' Ministry of Agriculture.

Republic of Kenya (2007). 'Kenya vision 2030.' Ministry of Planning and National Development and the National Economic and Social Council, Office of the President.

Republic of Kenya (2008). 'Economic review of agriculture.' Ministry of Agriculture, Livestock and Fisheries, Nairobi.

Ricker-Gilbert, J. and T. S. Jayne (2017). 'Estimating the enduring effects of fertiliser subsidies on commercial fertiliser demand and maize production: panel data evidence from Malawi.' *Journal of Agricultural Economics* 68(1): 70–97.

Ricker-Gilbert, J., C. Jumbe, and J. Chamberlin (2014). 'How does population density influence agricultural intensification and productivity? Evidence from Malawi.' *Food Policy* 48: 114–28.

Ricker-Gilbert, J., N. M. Mason, T. S. Jayne, F. Darko, and S. Tembo (2013). 'What are the effects of input subsidies on maize prices? Evidence from Malawi and Zambia.' *Agricultural Economics* 44(6): 671–86.

Rijkers, B. and R. Costa (2012). 'Gender and rural non-farm entrepreneurship.' *World Development* 40(12): 2411–26.

Ritzema, R., R. Frelat, S. Douxchamps, S. Silvestri, M. Rufino, M. Herrero, K. Giller, S. Lopez-Ridaura, N. Teufel, and B. Paul (2016). 'A simple food availability analysis across smallholder farming systems from East and West Africa: is production intensification likely to make farm households food-adequate?' Submitted to *Food Security*.

Rosegrant, M. W. (2008). 'Biofuels and grain prices: impacts and policy responses.' International Food Policy Research Institute, Washington, DC.

Rosegrant, M. W. and P. Hazell (2000). *Transforming the rural Asian economy: the unfinished revolution*. Hong Kong, Oxford University Press.

Ruben, R. (2001). 'Nonfarm employment and poverty alleviation of rural farm households in Honduras.' *World Development* 29(3): 549–60.

Rubin, D. and C. Manfre (2014). 'Promoting gender-equitable agricultural value chains: issues, opportunities, and next steps.' In *Gender in agriculture*. A. Quisumbing, R. Meinzen-Dick, T. Raney, A. Croppenstedt, J. Behrman, and A. Peterman. Dordrecht, Springer.

Sachs, C. (1996). *Gendered fields: rural women, agriculture, and environment*. Boulder, CO, Westview Press.

Sachs, J. D. and A. M. Warner (1997). 'Sources of slow growth in African economies.' *Journal of African economies* 6(3): 335–76.

Sakurai, T. and T. Reardon (1997). 'Potential demand for drought insurance in Burkina Faso and its determinants.' *American Journal of Agricultural Economics* 79(4): 1193–207.

Savadogo, K., T. Reardon, and K. Pietola (1994). 'Farm productivity in Burkina Faso: effects of animal traction and nonfarm income.' *American Journal of Agricultural Economics* 76(3): 608–12.

Schneider, K. and M. K. Gugerty (2010). 'Gender and contract farming in sub-Saharan Africa: literature review.' Evans School of Public Affairs, University of Washington.

Scoones, I. (1998). 'Sustainable rural livelihoods: a framework for analysis.' IDS Working Paper 72. Brighton, Institute of Development Studies, University of Sussex.

Scoones, I. (2009). 'Livelihoods, perspectives and rural development.' *Journal of Peasant Studies* 36(1): 171–96.

Sen, A. K. (1990). 'Gender and cooperative conflict.' In *Persistent Inequalities*. I. Tinker. New York, Oxford University Press.

Sheahan, M. and C. B. Barrett (2014). 'Understanding the agricultural input landscape in sub-Saharan Africa: recent plot, household, and community-level evidence.' Policy Research Working Paper 7014, Washington, DC, World Bank.

Shivji, I. G. (2006). 'The silences in the NGO discourse: the role and future of NGOs in Africa.' *Africa Development* 31(4): 22–51.

Simtowe, F. P. (2010). 'Livelihoods diversification and gender in Malawi.' *African Journal of Agricultural Research* 5(3): 204–16.

Simtowe, F. P. (2015). 'An assessment of national fertilizer policies, regulations and standards for Mozambique.' African Fertilizer and Agribusiness Partnership, AFAP.

Sitko, N. J. and T. Jayne (2014a). 'Exploitative briefcase businessmen, parasites, and other myths and legends: assembly traders and the performance of maize markets in Eastern and Southern Africa.' *World Development* 54: 56–67.

Sitko, N. J. and T. Jayne (2014b). 'Structural transformation or elite land capture? The growth of "emergent" farmers in Zambia.' *Food Policy* 48: 194–202.

Slavchevska, V. (2015). 'Gender differences in agricultural productivity: the case of Tanzania.' *Agricultural Economics* 46(3): 335–55.

Smale, M., D. Byerlee, and T. S. Jayne (2011). 'Maize revolutions in sub-Saharan Africa.' Working Paper Series 40/2011. Tegemeo Institute of Agricultural Policy and Development, Egerton University.

Snjiders, T. and R. Bosker (1999). *Multilevel analysis: an introduction to basic and advanced multilevel modeling*. London, Sage.

Sørensen, P. (1996). 'Commercialization of food crops in Busoga, Uganda, and the renegotiation of gender.' *Gender & Society* 10(5): 608–28.

Stockbridge, M. (2006). 'All Africa review of experiences with commercial agriculture: social impacts.' Background paper for the Competitive Commercial Agriculture in Sub-Saharan Africa (CCAA) Study. The World Bank. Available at: <http://siteresources.worldbank.org/INTAFRICA/Resources/257994-1215457178567/Ch13_Environmental.pdf>.

Thomas, S. (1975). 'Economic developments in Malawi since independence.' *Journal of Southern African Studies* 2(1): 30–51.

Timmer, P. (2009). *A world without agriculture: the structural transformation process in historical perspective*. Washington, DC, American Enterprise Institute Press.

Timmer, P. and S. Akkus (2008). 'The structural transformation as a pathway out of poverty: analytics, empirics and politics.' Working Paper 150. Center for Global Development, Washington, DC.

Tittonell, P. and K. E. Giller (2013). 'When yield gaps are poverty traps: the paradigm of ecological intensification in African smallholder agriculture.' *Field Crops Research* 143: 76–90.

Udry, C. (1996). 'Agricultural production and the theory of the household.' *Journal of Political Economy* 104(5): 1010–46.

Udry, C., J. Hoddinott, H. Alderman, and L. Haddad (1995). 'Gender differentials in farm productivity: implications for household efficiency and agricultural policy.' *Food Policy* 20(5): 407–23.

United Nations (2014). 'World urbanization prospects: the 2014 revision.' Department of Economic and Social Affairs, Population Division, New York.

United Republic of Tanzania (2008). 'Agricultural Sector Development Programme (ASDP): third joint implementation review, an assessment of the agricultural services (research and extension) delivery under ASDP.'

United Republic of Tanzania (2015). 'Agriculture Sector Expenditure Review, Tanzania mainland.' Dar es Salaam.

United Republic of Tanzania, Ministry of Finance and Economic Affairs (2010). 'National Strategy for Growth and Reduction of Poverty II (NSGRP II).'

United Republic of Tanzania, National Bureau of Statistics (2006). 'National account of Tanzania: mainland 1992–2004.' Dar es Salaam.

USDA (2016). 'World agricultural production.' Circular Series WAP 9–16, September 2016. US Department of Agriculture Foreign Agricultural Service/Office of Global Analysis.

Valdés, A. and W. Foster (2007). 'Making the labor market a way out of rural poverty: rural and agricultural labor markets in Latin America and the Caribbean.' Background paper for the *World Development Report 2008*. Washington, DC, World Bank.

Van Donge, J. K. (2002). 'Disordering the market: the liberalisation of burley tobacco in Malawi in the 1990s.' *Journal of Southern African Studies* 28(1): 89–115.

Vella, F. (1993). 'A simple estimator for simultaneous models with censored endogenous regressors.' *International Economic Review* 34(2): 441–57.

Von Braun, J. and R. Pachauri (2006). 'The promises and challenges of biofuels for the poor in developing countries: IFPRI 2005–2006.' Annual Report Essay, International Food Policy Research Institute, Washington, DC.

Von Braun, J. and P. J. Webb (1989). 'The impact of new crop technology on the agricultural division of labor in a West African setting.' *Economic Development and Cultural Change* 37(3): 513–34.

Walker, C. (2003). 'Piety in the sky? Gender policy and land reform in South Africa.' *Journal of Agrarian Change* 3(1–2): 113–48.

Wambugu, S. K. and J. T. Karugia (2014). 'Poverty impacts of straddling: rural income diversification in Nyeri and Kakamega counties, Kenya.' Afrint III report.

Whitehead, A. and N. Kabeer (2001). 'Living with uncertainty: gender livelihoods and pro-poor growth in rural sub-Saharan Africa.' IDS Working Papers, 134. Institute of Development Studies, Brighton.

Whiteside, M. (2000). 'Ganyu labour in Malawi and its implications for livelihood security interventions: an analysis of recent literature and implications for poverty alleviation.' Overseas Development Institute, London.

Wiggins, S. (2000). 'Interpreting changes from the 1970s to the 1990s in African agriculture through village studies.' *World Development* 28(4): 631–62.

Wiggins, S., G. Argwings-Kodhek, J. Leavy, and C. Poulton (2011). 'Small farm commercialisation in Africa: reviewing the issues.' Research Paper, Future Agricultures Consortium.

Wiggins, S. and S. Keats (2013). 'Leaping and learning: linking smallholders to markets in Africa.' Agriculture for Impact, London, Imperial College.

Wiggins, S., S. Keats, and J. Sumberg (2015). 'Agricultural growth in the New Alliance countries.' Future Agricultures Working Paper 108.

Wooldridge, J. M. (2002). *Econometric analysis of cross section and panel data*. Cambridge, MA, MIT Press.

World Bank (1994). 'Tanzania agriculture.' Washington, DC, World Bank.

World Bank (2007). *World development report 2008: agriculture for development*. Washington, DC, World Bank.

World Bank (2015). 'Food price watch.' Washington, DC, World Bank.

World Bank (2016). 'World development indicators.' Washington, DC, World Bank.

World Bank, FAO, and IFAD (2009). *Gender in agriculture sourcebook*. Agriculture and Rural Development Series. Washington, DC, World Bank.

Yaro, J. A. (2006). 'Is deagrarianisation real? A study of livelihood activities in rural northern Ghana.' *Journal of Modern African Studies* 44(1): 125–56.

Zant, W. (2012). 'Is EU support to Malawi agriculture effective?' Tinbergen Institute Discussion Paper.

Zwarteveen, M. (2008). 'Men, masculinities and water powers in irrigation.' *Water Alternatives* 1(1): 111.

Index

The manufacturer's authorised representative in the EU for product safety is Oxford University Press España S.A. of el Parque Empresarial San Fernando de Henares, Avenida de Castilla, 2 – 28830 Madrid (www.oup.es/en or product.safety@oup.com). OUP España S.A. also acts as importer into Spain of products made by the manufacturer.

www.ingramcontent.com/pod-product-compliance
Lightning Source LLC
LaVergne TN
LVHW011751200926
848484LV00003B/674
9780198799283